FIRE ENGINES

Hans Halberstadt

MBI Publishing Company

*For my dear wife April, who has been holding out for a book
with a lot more sex and violence in it—this will have to do.*

This edition published in 2002 by MBI Publishing
Company, Galtier Plaza, Suite 200, 380 Jackson
Street, St. Paul, MN 55101-3885 USA

The information in this book is true and complete
to the best of our knowledge. All recommendations
are made without any guarantee on the part of the
author or Publisher, who also disclaim any liability
incurred in connection with the use of this data or
specific details.

We recognize that some words, model names and
designations, for example, mentioned herein are the
property of the trademark holder. We use them for
identification purposes only. This is not an official
publication.

MBI Publishing Company books are also available
at discounts in bulk quantity for industrial or sales-
promotional use. For details write to Special Sales
Manager at Motorbooks International Wholesalers
& Distributors, Galtier Plaza, Suite 200, 380
Jackson Street, St. Paul, MN 55101-3885 USA.

Library of Congress Cataloging-in-Publication Data
Available

ISBN 0-7603-1366-0

On the front cover: This beautiful 1963 American
LaFrance pumper was purchased right out of
service for $2,000 in 1992 in virtually perfect
condition.

On the frontispiece: This 1931 Mack is operated by
the San Jose Fire Department's Muster Team.

On the title page: San Jose's 1937 American
LaFrance Metropolitan. Ownership of such a rig
can produce all kinds of dilemmas when, for
example, choices have to be made about repainting
a weathered and cracked finish that also contains
gold leaf and pinstriping superior to anything that
could be applied today.

On the back cover: Mack has long been popular
with the Fire Department of New York, with
hundreds of sales over the years.

Printed in Hong Kong

Contents

	Acknowledgments	7
Introduction	Heroes on Wheels	9
Chapter 1	Engine 8, First In!	13
Chapter 2	Fire Fundamentals	29
Chapter 3	Vollies and Steam	41
Chapter 4	The Beginning of the Automotive Age	67
Chapter 5	Collectible Fire Apparatus Manufacturers Histories	129
Chapter 6	Modern Apparatus	171
Chapter 7	Overhaul and Take-up	187
	Appendices	190
	Index	192

Acknowledgments

Writing any book is a team sport. I've been nobly assisted on this project by an all-star assembly of players and coaches from fire departments, associations, museums, and collector groups. In particular, the San Jose Fire Department has been tolerant and gracious about having me under foot. Capt. Dennis Madigan has been extremely supportive, but so have the firefighters of many stations, including Firefighter Juan Diaz, Capt. John Charcho, Deputy Chief Ed Hart, Capt. Henry DeGroot, the entire A-shift staff at Station 8—Capt. Dick Toledo, Firefighter Tony Magallon, Firefighter Gary Zobrosky, Fire Engineer Al Souza—and many others.

Much of the inspiration for the project came from California's Central Fire Protection District's Education Officer Christie LeBaudour and that department's excellent cast of characters. On occasion, they have actually paid me to photograph their rigs and their people.

The fire department in Campbell, California, posed gleefully, shuffled rigs, and tolerated an alien aboard for ride-along duties.

Many private owners of trucks and engines helped me by offering time, advice, and the use of their precious vehicles. Ed Hass, president of the Ahrens-Fox Fire Buffs' Association, provided generous access to his marvelous library and artifacts collection. Many others contributed including Len Williams, Zoltan Szucs, Cliff Smith, Steve Flarty, Kenneth Betchtel, Bob Ward, president of the Society for the Preservation & Appreciation of Antique Motor Fire Apparatus in America, (SPAAMFAA), and Dana Martin.

Sara Saetre has worked her usual editorial magic, providing insights and counsel on the form and content of yet another disorderly set of scribblings.

Retired firefighters Lt. Pat Madigan and Fire Engineer Loren Gray provided wonderful insights into the old days, as did Battalion Chief and Master Mechanic (retired) Don Wisinski with his assistance on the new rigs.

Dr. Peter Malloy, director of the Hall of Flame, contributed most of an entire week to this book. To assist this project, he provided access to the entire resources of the world's largest fire-safety museum.

To all hands, I'm extremely grateful.

During the transition from hand- and horse-drawn apparatus to motorized apparatus, the old- and new-style vehicles had much in common. A 1924 American LaFrance Type 40 chemical and hose car rolled to work for many years on these wooden wheels with solid-rubber tires. This vehicle is part of the Hall of Flame collection.

Heroes on Wheels

I must have been about five years old the first time I really noticed that big, red, loud, fire engine as it went roaring down the street. Like just about every other kid, well—it was love at first sight! They say that first loves are true loves, and that is certainly the case for many people when it comes to fire engines. Many of us can remember their first with all the exquisite detail that we also associate with members of the opposite sex. Some of us have made fire engines (and trucks, too) a kind of life obsession.

Fire trucks and engines are special. They have always had an important role in US and Canadian society, from the Colonial days to the present. They are weapons in a heroic struggle against an ever-present menace, and they have a glitter and glory that even military aircraft can't quite match. Fire engines—from the hand pumpers of 200 years ago to the huge monster crash trucks used today at airports—are amazingly varied in size and appearance, but nearly all are in some way beautiful and attractive. There are very, very few ugly fire engines from any place or any time in North America.

This book is a portrait of fire apparatus—it is not only about the fire engines but of the trucks, ladder rigs, crash trucks, and other vehicles that work on the fire ground. It is an unusual portrait because I've combined the stories of particular engines or trucks with the jobs they do fighting fires. Firefighting has changed over the years, and the apparatus used to fight fires has changed with the evolution of our cities and societies.

This book's title, *The American Fire Engine*, includes not only fire apparatus in the United States but also in Canada. Our cultures are and have always been kissing cousins, and the development of technology in the United States has been invariably linked to our northern neighbors. After all, the first motorized rigs that the Seagrave Company sold went to Vancouver, British Columbia. The links between departments and manufacturers on both sides of the border have always been strong and friendly.

Some fire buffs may object to my use of the word "engine" in the title of a book that is about all kinds of fire apparatus—trucks, engines, aerial ladders, and the odd snorkel and super pumper. And in a way they are right because, to collectors and firefighters, a fire *truck* is a different beast than a *fire engine*—but the general public doesn't make the distinction. This book is the story of vehicles that fight fires in the cities and towns of Canada and the United States, whether a hand pumper from the 1850s or the huge Oshkosh T-3000 airport crash truck of today.

There are two major groups of fire enthusiasts: the fire buffs (or "sparks" as they are sometimes called in parts of the Northeast) who are fascinated by the working fires, and the apparatus buffs who enjoy the technology of the vehicles. These two groups have their own organizations (some very specialized) and they occasionally speak to each other despite radically different points of view on fire apparatus. The fire buffs don't particularly care

Fire departments cherish their old trucks and engines, keeping them polished and running for decades after the rigs were retired. The Morgan Hill, California, Fire Department has owned this beautiful 1926 Seagrave Suburbanite since it was new.

about the specifications of the rigs that go to a fire. They would not object if a burning wall happened to fall on one; if anything, that would add to the satisfaction of the event. The other group, the vehicle fans, would really rather the trucks and engines never left the fire station—because they might get dirty or scratched somehow, and that would be awful.

Somewhere in between these two communities are the firefighters themselves—who normally love a good burner (as long as nobody gets hurt), but who also admire the apparatus of the present and adore the rigs of the past. And since they not only fight the fires but also clean up the rigs afterwards—well, I'm with them.

Firefighting is still the most hazardous occupation in our society. It is a stressful business for the firefighters themselves and for the vehicles that get them to the fires (and car wrecks, chemical spills, medical calls, false alarms, and other disasters). Fire engines last about twenty years in service; firefighters last about thirty before they retire. And while firefighters statistically don't live as long as the general population, the trucks and engines they ride seem to be immortal and remain healthy for—in many cases—nearly 200 years. Both have played a tremendous role in our culture, and this is the story of both, together, on what firefighters call the fire ground, the area immediately adjacent to an incident.

Over the last few years, I have worked extensively with the US Army's Green Berets, the Navy's SEALs, the Marine Corps, and the Air Force's F-15 fighter pilots. I have studied the communities, missions, and people within. The bravery and skill of these people is unquestionable. After working with the men and women of the fire service, however, I discovered that the average American or Canadian firefighter, during the course of a career, confronts more death, destruction, and personal hazard than even those marvelous friends of mine in the military services. Relatively few people in the military ever hear a shot fired in anger, ever see death face to face, ever experience the primal fear of extreme danger. But many people in the fire service confront death with great frequency, mangled bodies fairly often, and extreme personal risk many times.

The Fire Engine as a Weapon

A fire engine is, and has been for all of American history, a wonderful, benign weapon used to fight battles against the old enemy, the red devil, fire. When a first-in fire engine company rolls up to a building that is fully involved (the firefighter's phrase for a place with flames showing and people known to be inside), they conduct an interior attack. A door or window is forced. With the line charged with water, the nozzle set, and the engineer operating the engine pump panel, the hose team advances into the burning building.

Flames may be running the ceiling, dancing with an eerie, liquid grace. The air inside is much like being inside a kitchen oven when the thermostat is set for 350 degrees. The hose team crawls, slides, or duck-waddles forward toward the flames behind a fog of life-sustaining water. If the engineer, outside at the pump panel, should fail to supply water now, the firefighters advancing the line will suddenly be exposed to the full effects of the flames, and will almost certainly be added to the long list of injured or killed firefighters for the year.

When the firefighters reach the seat of the fire, its heart and foundation, the nozzle pattern is changed; the firefighter on the nob sweeps the nozzle in a fast, clockwise pattern, knocking down the fire—whose heat converts the water to steam, engulfing the firefighters in a stifling, searing fog. Even the nozzle itself is a threat to the firefighter. If it slips from a firefighter's grasp (and it can), the 100 to 250 gallons per minute of water rushing out the tip will turn the hose into a wildly thrashing, deadly snake capable of striking a killing blow (which has happened). In this hot, dank, deadly atmosphere, firefighters search for survivors or victims, often with sad success. It is part of the job to discover limp babies and the charred bodies of the dead or soon-to-be-dead. Interior attack can sometimes be like going inside an oven with the roast.

Such fires, however, are a small part of a fire company's duty. But the routine work of emergency medical calls (usually about 80 percent of the responses of an engine or truck company) require considerable bravery, skill, and judgment. Extracting mangled people from car wrecks, performing CPR and medical procedures on HIV-positive patients, trying to force life back into people who have just had heart attacks, strokes, or gunshot wounds—they are all part of the shift duty for almost any urban professional department or the many volunteer companies around the United States and Canada. One volunteer fire-fighter I know has done CPR on over 50 dying people—watching them drift slowly away despite his desperate attempts to keep the blood moving in the arteries, the air moving in the lungs. Of all these people, only one survived, but even so he says it is worth the effort.

Although this is a book mostly about the big, beautiful vehicles we call fire engines, you can't really understand fire *apparatus* without a basic understanding of the fire *service*—its missions, techniques, heritage, and traditions. Those things all shape the engines, trucks, and ladder rigs in subtle ways, which end up in sheet metal and in the layout and design of hose beds, cabs, engines, equipment compartments, drive trains, and even paint schemes. The missions, heritage, and traditions also shape the people who ride the rigs, producing men and women with a special kind of courage and commitment and a gentle generosity unique to the fire service. These rigs, and the men and women who use them, are all heroes on wheels.

S an Jose's 1931 Mack has been a remarkably durable rig. After sixty years it is still running strong and still serves the department. When retired firefighters die their caskets are often borne to the cemetery in the hose bed of this engine. It has outlived most of the men who fought fires with it.

Chapter 1

Engine 8, First In!

There is an old Catholic church about a mile from where I live—one of those big, ornate, neo-Spanish colonial structures that was a popular style of church architecture around 1910. It is called Five Wounds, and for eighty years has been the center of the Portuguese community in our area. People come from many miles for weddings, funerals, christenings, and Sunday services. The building has become a fixture in the traditions of many local families.

As the church served the community, the I.E.S. (Irmandade do Espirito Santo) hall next door served the church. The hall was the place for the reception after the wedding, for the band to practice and store their instruments and uniforms, for gatherings after church, and for bingo on Wednesday evenings. The hall had meeting rooms, a kitchen, trophies, and memories. It was a sweet building, old and worn, polished with use and the presence of many generations. The church still stands but the charred remains of the hall were bulldozed during November of 1992, shoveled into trucks and delivered to a landfill—a mixture of burned wood, electrical wiring, mangled band instruments and religious statues, charred sheet music, and a few melted items of firefighting apparatus that were abandoned to civilization's ancient enemy, fire.

The Five Wounds hall had been through many changes over the decades, each change made with the best of intentions. About fifty years ago, the hall got an additional single story built on the back. Air conditioning units were installed sometime in the distant past. Partitions were added, along with false ceilings. Each change adapted the building to the people who used it.

A couple of years ago, the community started a new renovation of the hall. The church raised $700,000 to invest in improvements. People donated time and effort, in addition to money. The renovation took a long time to complete, but finally the project was about done. The parishioners planned a big party to celebrate its renewal.

But on a hot Wednesday afternoon in August, the wiring for an air conditioner got crossed and a short circuit developed. Somehow, the circuit breaker that was supposed to protect the building failed to trip. The wires got hot and the material near the unit slowly warmed until it ignited. Twenty parishioners and a priest, Father Charles Mecado, were setting up for a bingo night in the auditorium portion of the hall when the lights flickered, then went out. Somebody noticed the smell of smoke. Father Mecado called 911 while everybody else evacuated the building.

The call comes in to Fire Communications at 16:55:55 (just before 5:00 pm): "Smoke is coming from the roof and the lights have gone out. Could the fire department send somebody out to take a look?" The call, like the hundreds that come in every day, was entered into the computer at 16:56:44.

About a mile from the hall, at the corner of 17th and Santa Clara in the heart of downtown is

Flashover! This is "truckie" work, opening up the building to vent the hot, poisonous, explosive gasses. The pike pole being used is virtually unchanged from the tool used for the same purpose hundreds of years ago. When a sealed building begins to burn, the available oxygen is quickly consumed and the fire may appear to have died down. All it takes to have it blow up in your face is to carelessly open a door, or have a window break, letting fresh oxygen feed the flames. Then, no matter how hot the day, you want to be dressed in heavy, insulated Nomex pants and coat, with protective gloves, mask, helmet and hood, with a Scott air pack feeding you fresh air. It is work like this that makes firefighting the most hazardous of American occupations.

A heavy "master stream" is directed onto a fire from high above the ground. The business of fighting fires shapes the specifications for fire trucks and engines, changes their shapes and sizes, function determining form.

Station 8, one of the oldest in town. It used to be Chemical 6 and was—back in the 1920s—on the outskirts of town. The chemical wagon has long since been replaced by four generations of automotive rigs, but the building is little changed from seventy-five years ago. There is room for only one rig in the station and currently in residence is a sturdy 1988 Pierce-Arrow pumper with a 1,500 gallons-of-water-per-minute (gpm) capacity. There is 700 feet of 5-inch supply line, 1,400 feet of 3-inch hose, 200 feet of 1 3/4-inch preconnected hose, 150 feet of 1 1/2-inch preconnect and 150 feet of 2 1/2-inch preconnect. The engine, and the people who use it, are collectively known as

Engine 8. Four people crew Engine 8: the captain, an engineer, and two firefighters.

You know the legend about the great meals served in fire stations? Well, at Station 8 you might get a baloney sandwich for lunch—if you're lucky to get lunch at all. Station 8's rig is in and out of the place so often during a shift that there is hardly enough time to heat a can of soup from one call to the next. The folks at Station 8 make an attempt at a traditional firehouse dinner, but have long since given up trying elaborate recipes that take extensive preparation or that cannot be interrupted. But even so, on this August afternoon Fire-fighter Gary Zobrosky the resident chef on the A shift, assembles his Chinese chicken salad in the kitchen while the other three crew members recover from the previous call.

Most of the runs Station 8 receives are EMS (emergency medical services) calls in addition to the occasional vehicle lockout, car wreck, false alarm, and floating corpse in the creek. Every day brings suicide attempts, successful and otherwise, heart attacks, drug overdoses, and scraped knees on four-year-olds. Gone are the days when the Station 8 crew would get your cat out of a tree, but just about any other emergency (real or imagined) is fair game. Once in awhile, Station 8 even gets to

Firehouse Cooking

Firehouses are famous for their cooking, and with good reason; there is a long tradition of hearty, sometimes sophisticated cuisine in many fire departments. A fire company is often just the right size to shop and cook for, from four to ten, and in most houses today everybody chips in for the groceries. One or two firefighters who like the job (and do it well) cook the meals. Afterwards, the crew will sometimes roll dice to decide who washes up.

There are dozens of great recipe books of firehouse cooking, and almost any bookstore will stock a few. Many departments publish their own and sell them at fund-raisers.

Many restaurants around the country specialize in what they allege to be firehouse cuisine, some because they are located in old firehouses, others because the owners are current or former firefighters, still others because they couldn't think of a better theme. *The Visiting Fireman* (see Recommended Reading) lists many of these places and provides a thumbnail review of each.

As a good sample of the genre, here is the recipe for Patented, Certified Wholesome Station 8 Meat Loaf, as prepared by Firefighter Gary Zobrosky, San Jose's own A-shift chef.

Station 8 Meat Loaf

2 eggs
4 pounds lean ground chuck
2 onions, chopped
1 bell pepper, chopped
2 packages commercial meat loaf spice mix

Preheat oven to 350 degrees. Mix meat and eggs in a large bowl. Add the remaining ingredients and mix with your fingers. When the mixture is ready and stuck to fingers, receive call to respond to a car wreck on the freeway. Place bowl in refrigerator, wash hands, don't dry, put on turn-outs, and go on call. Return in about 45 minutes and clean engine.

Wash hands again and take the mixture from refrigerator. Mold into a mound (don't place in conventional bread pan) and place in aluminum foil-lined flat baking pan. Place in oven and begin to cook for approximately 15 minutes before getting call for an EMS response. The oven will shut off automatically when the call comes in; return in 30 minutes.

Reset oven shutoff, reheat oven, and allow to cook for 5 minutes before getting call for an alarm pulled at a middle school; go to call (false alarm, of course) and come back 20 minutes later.

Add large baking potatoes to the oven and monitor progress of meat loaf. After 45 minutes, remove the meat loaf and potatoes; total baking time for meat loaf should be 1 1/2 hours.

Get the captain to make his famous salad and set the table. Sit down to dinner and get another call, this time for a stove fire; put the potatoes and meat loaf back in the oven to keep warm. Come back in 1 hour and have dinner—with any luck there won't be any interruptions. Serve with coffee, milk, or soda. *Estupendo!*

Engine 8, operating from the west side of the I.E.S. (Irmandade do Espirito Santo) hall, makes its contribution to the effort to save the structure. Firefighter Gary Zobrosky operates the big deck gun on the top of the rig, adding a master stream of over 750gpm to the flames. Over a million gallons of water will be sprayed on the fire—and it won't be enough.

CLARA STREET . . . ONE FOUR ZERO ONE EAST SANTA CLARA STREET . . . BETWEEN 28TH AND 30TH STREETS. STATION 8'S AREA—ENGINE 8, FIRST DUE.

The dispatcher sends the standard "structure response": two engines, one truck, and a battalion chief. Because the call came from Station 8's primary area, they are closest and assumed to be "first due." Three stations hear the assignment come over the system, including Station 1 (a large firehouse downtown) and Station 10 (5 miles and about 12 minutes from the incident). The call came in while most of the complement from Station 1 was responding to another fire. Normally, the dispatcher would assign Battalion 1 (the battalion chief from Station 1), but he is already busy, so Battalion 10 (Battalion Chief Jeff Clet) gets this call, even though he is miles away.

Even before the diesel engines start and the doors open, all the players know the routine and their individual role in it. Station 8's temporary boss Capt. Paul King, will be in charge—the IC (Incident Commander)—until Battalion Chief Jeff Clet (Battalion 10 in the dispatch) relieves him. Engine 8's responsibility as first in is to evaluate the situation ("size up" in the trade). The partially assembled chicken salad goes into the refrigerator. As King, Fire Engineer Al Souza (who drives the rig and operates the pump panel), and Firefighters Zobrosky and Tony Magallon walk quickly toward the engine, they know their roles. What they don't know is that it will be a long, hot night, and that dinner will have to wait until 02:30 the next morning.

The firefighters walk, not run, to their rig. The turn-outs wait just where they like to have them, on or beside their positions on the vehicle. Their regular work boots have zippers on them and come off almost instantly. They step into the big rubber fire boots and pull on the turn-out pants over their work pants. They shrug into the suspenders and pull on the heavy turn-out coat. Even though it is a hot August afternoon, they clip the four snaps. Now they put on their helmets, with plastic face shield, and snug up the chin straps. The firefighters climb aboard and strap in—the whole process takes 15 seconds. In Station 8's department, firefighters also strap into self-contained breathing apparatus (SCBA) when

Turning left out of Station 8, Fire Engineer Al Souza takes the rig up the street toward the site of the seven-alarm fire. Modern rigs like this 1989 Pierce are configured more like fighter airplanes than like conventional trucks—and cost about as much as an aircraft as well.

go to a fire. They average ten calls every 24-hour shift in the busiest station in the city in one of the busiest firehouses in the nation.

At 16:56:44 on that hot August afternoon, the station gong strikes once, as it does for every assignment, a loud, clear *bong* that firefighters can hear anywhere in the building. The gong warns that an assignment is about to ring down; everybody stops, waits, and listens. The teleprinter starts to tap, printing out the details of the call. Over the station loudspeaker comes the calm, careful voice of the dispatcher in the Fire Communications Center downtown. She says,

STATION 8 . . . ENGINE 101 . . . TRUCK AND RESCUE 1 . . . BATTALION 10 . . . A STRUCTURE RESPONSE, 1401 EAST SANTA

they climb on the rig. Everyone is ready for battle by the time the rig rolls onto the scene.

Captain King tears off the top copy of the printout on his way to his position on the rig. He climbs up into the right seat and straps in. In front of him is a detailed map of Station 8's area, which includes the location of hydrants. King studies it carefully.

The station door opens. Souza doesn't have to worry about finding a key—there isn't one. He sets the battery switch to BOTH, the ignition switch to ON, and presses the two start switches momentarily. The huge diesel rumbles instantly to life.

Souza puts his foot on the brake, moves the transmission selector from PARK to DRIVE and eases out of the station. (The new trucks all have automatic transmissions and power steering, just like a car that weighs 10,000 pounds .)

The report of a structure fire rates a "code three" response: lights and siren. The light switch is energized and the light bar on top of the rig flashes like a Las Vegas casino sign. Firefighter drivers can drive fast and go through red lights with a fire engine, but they must know how to drive a big, heavy truck fast, and they had better go through red lights at intersections carefully, with

The "house-wreckers" from Truck 1 get to work on the roof of the IES hall. Below them, in the hidden spaces between the roof and the interior ceiling, fire has already infected the structure. Even with the roof going hot and soft under their boots, they have the job of ventilating the building by cutting holes in the roof.

It's hard to have a seven-alarm fire without having the news media show up and get under foot. That's Truck 1, with its 85-foot snorkel, doing the honors at the rear of the building and adjacent to the seat of the fire. The red helmet identifies the firefighter as a captain—who normally doesn't do this kind of work. Despite all the advances in technology, the stream he's directing is about the same volume of water as that produced by fire engines of 150 years ago—mostly because hoses of higher power are too difficult for one person to control.

plenty of air horn and a careful look around the corner before proceeding.

Capt. King taps the siren button on the floor with his foot and Souza leans on the air horn as Engine 8 rolls out the door and onto the street. Traffic usually stops—but not always. Fire Engineer William Anger was killed a few years ago when a drunk driver didn't notice the lights or siren and plowed into the rig at an intersection. Station 8's firefighters are still extra careful. Engine 8 rolls out onto the street and turns a hard right up Santa Clara Street toward the Five Wounds hall, with a blast of the air horn, lights flashing and siren wailing.

Size-up begins long before the engine rolls onto the fire ground. King can see the area as soon as he is out on the street, a straight shot down Santa Clara; little smoke is visible and no flames. At 17:00:04, Engine 8 rolls up to the hall, stopping on the east side of the structure. As Souza dri-

ves the engine into the parking lot, there is still little smoke showing—until they pull around on the east side of the structure where moderate amounts of gray smoke pour from the eaves of the rear portion. In front of the building, Father Mecado and about twenty parishioners wait anxiously. Magallon and Zobrosky pull the 1 3/4-inch line off the engine and dive into the building, continuing the size-up of the incident. They walk inside, leaving the dry hose at the door, looking for a fire to fight.

Deep in the structure, back from the street, there is smoke but no flame. Puffs of smoke come from the air conditioner vents along the ceiling. With pike poles, Magallon and Zobrosky pull out sections of tiles—and more smoke gushes out. But the spaces are generally clear; there seems to be a problem, but just where and how bad are not immediately apparent.

King orders, "Charge the line!" and water flows into the dry hose, toward the nozzle. Souza charges the line with water from the tank aboard the rig—500 gallons—enough for about three to five minutes of delivery. Engine 8 is ready to "tank" the fire, but isn't yet tied into a hydrant. If this turns into a real burner, King is going to need lots of help. That help is on the way aboard Engine 101 from the downtown Station 1, a mile from the scene.

The captain calls the Fire Communications Center and reports, ON SCENE. . . SMOKE IN AREA . . . INVESTIGATING.

Captain King calls again moments later at 17:01: IT'S A LARGE BUILDING . . . POTEN-TIAL INVOLVEMENT TO THE REAR. SEC-OND ALARM! WHO'S MY SECOND-IN COMPANY?

Engine 101 rolls up to the scene, hears the transmission, and calls King directly on the main fire channel. 101! WE'RE OUT AT THE HYDRANT! Capt. John Charcho informs King. Engine 101, as second-in has the responsibility to run a supply line from the nearest hydrant to Engine 8's intake, a big 5-inch hose that has to cross the busy boulevard out front. Charcho calls the dispatcher with a demand for the police to shut down the street: GET THE PD OUT HERE TO CONTROL TRAFFIC, ASAP! WE'VE GOT TO PUT A 5-INCH ACROSS SANTA CLARA!

The 5-inch is the biggest hose a fire engine carries. It essentially moves the fire hydrant to the fire engine, providing all the water volume and pressure that the hydrant has to offer without the pressure and volume loss that results with smaller hoses. Until the 5-inch is hooked up and "charged," the cast of characters from Engine 8 are restricted to what is in their on-board booster tank. But the hydrant is across the street from the fire. The big hose will totally shut down a major thoroughfare, right at rush hour, so the business of getting water to the fire in volume is momentarily complicated.

Truck and Rescue 1, another company from the downtown station, arrive and proceed to work their magic. The truck company "ladders" the back portion of the structure and clambers up to inspect and begin ventilation. Battalion 10 (Jeff Clet) is en route from his distant station, but stuck in rush hour traffic on the freeway; the party proceeds without him. Engine and truck

crews have specific duties on the fire ground and now, while the engine crew attacks the fire, the truck crews make a thorough search of the building for anybody still inside and begin ventilating the building.

Ventilation is a scary, dangerous business. It requires getting up on the roof of a building and chopping holes in it while the fire eats away at the beams and support surfaces below. As firefighters ventilate a building, they can see the roof go soft in front of them—and if they aren't careful (and even if they are) the surface can suddenly collapse and they will be consumed by the red devil. But heat rises, and the hole makes the building breathe; ventilation cools the interior and cleans the air. It lets firefighters control the fire, stop it from spreading, and put water on it. The only way to do that is for crews to get on the roof and open up the structure.

Inside the structure things are going to hell in a hurry. King has only two people to find and physically attack the fire at this moment; he is supposed to supervise and use the radio to command and coordinate. But when they poke a hole in a ceiling and smoke and fire come out, King knows they need more people, fast. Engine 101 has to run the supply line from the hydrant before they can join in the interior attack and the second alarm units are just stepping into their turn-outs at their stations, miles away.

Truck 1 attacks the fire from overhead with a high-volume master stream. The great virtue of snorkels is the ability to put serious water into the seat of the fire from an angle that would be impossible any other way. But not even the snorkel can get water on the fire inside the ceiling of this building. The flames are showing from the area where the fire apparently began, in the wiring for an air conditioning unit hidden in the smoke.

Engine 101 backs up King and his crew with the 2 1/2-inch attack line from Engine 8's pump. King assigns Engine 101's Charcho to be Division A commander.

With the second alarm goes a routine group page summoning a small multitude of supporting players, including the department's volunteer photographer, me. My fire department voice pager, along with those of all the battalion commanders, chiefs, and assorted specialists, went off at one minute past five, the calm and authoritative voice of the dispatcher announcing: SECOND ALARM, STRUCTURE RESPONSE, 1401 EAST SANTA CLARA STREET. . .ONE FOUR ZERO ONE EAST SANTA CLARA STREET. . . BETWEEN 28TH AND 30TH.

I collect my gear and drive (as calmly as possible) to the fire, half a mile from my office. The smoke is visible as soon as I turn onto Santa Clara, black and heavy, rolling across the street up ahead.

I park a block away from the fire ground. It is going to be a crowded, congested place later and I don't want to be trapped. The police department is already on scene and has closed the street, access to the fire ground is controlled. On the street are the curious and the concerned—the first group gleeful and the latter, appearing to be parishioners, scared and shocked. I show a cop my pass and cross the fire line.

Engine 8 has its hands full. Fire has infected the building through the hidden spaces behind walls and ceilings. King, still the IC, is inside, trying to find the fire, but it is already everywhere in the attic spaces of the old building. Despite the modern equipment on the fire ground, the crews from Engine 8 and Truck 1 inside use a tool that is hundreds of years old—the pike pole—to pull away sections of the ceiling in an effort to expose the fire.

What they find is layers of ceilings, one hiding another, with the fire between the last one and the roof in a space firefighters call the "cockloft." King calls desperately for ventilation of the roof and more firefighters inside. The crew from Truck 1 is already on the roof, cutting holes in it with power saws, trying to get at the seat of the fire and to vent off the smoke, flame, and heat that are accumulating inside. Ten minutes after the hour, Battalion 10, Jeff Clet, arrives and assumes the role of IC from Engine 8. At twelve minutes past the hour, he keys his radio and calls, COMMAND 4, SANTA CLARA IC . . . DISPATCH A THIRD ALARM. Engine 2, Engine 16, Truck and Rescue 3, and Battalion 2 are assigned to the fire.

From inside, the Division A commander radios urgently, SANTA CLARA IC, DIVISION A INSIDE. . .WE GOT FIRE RUNNING THIS CEILING! GET THAT ROOF OPENED UP! WE NEED ANOTHER TRUCK CREW WITH PIKE POLES IN THE INTERIOR! Rescue 5 enters to help with the interior attack. On the roof, the struggle to cut a trench continues with K-12 saws and special carbide-tipped-blade chain saws.

It is a weird fire—inside, conducting the interior attack, the firefighters grope through inky smoke and intense heat but can't find the seat of the fire. When they pull off a section of wall with pike poles and hand tools, smoke and flame pour out at them. Out on the street the big 5-inch supply line is being run from a hydrant across the street to feed Engine 8. King calls for a second 2 1/2-inch line. Over his head the flames ripple and flow like a kind of hellish liquid. Outside, one firefighter, already soaked, exhausted, impregnated with black grime, turns to me and says, "We're going to lose it."

As the truck and engine companies arrive on scene, they stage two blocks away and receive their assignments from there. Engine 26 arrives and is dispatched to Exposure 4, ordered to drop a 5-inch supply line on their way in and then to assist Engine 8 with their hand lines. The big supply line essentially moves the hydrant to the engine, although it totally blocks traffic on any street on which it is laid. By the time the fire is out, these big lines will be running from every hydrant within a quarter mile.

Truck 1's captain, Vince Jangrus, calls, SANTA CLARA IC. . .TRUCK 1! THE HEAD OF THE FIRE IS NOW OUTSIDE! ALL THE HOLES WE'VE CUT IN THE ROOF ARE NOW VENTING FIRE. BE REAL CAREFUL WITH THE ROOF. . . I'VE GOT TWO GUYS ON THE ROOF RIGHT NOW!

ENGINE 8. . .WE NEED A HOLE ON THE SOUTH SIDE OF THIS ROOF AND SOME 1 1/2 OR 1/3/4 LINES INSIDE!

The glamorous work of the firefighter is often just pure brute strength. With a couple of hundred gallons of water squirting out of the "nob," the man on the nozzle must maintain a force of about 30 pounds to keep control. That may not seem like much until you've tried it for half an hour or so, even with some help from your friends. The actual business of firefighting, on an engine or a truck crew, requires plenty of upper-body strength, agility, and good judgment to go along with the skills they teach you at the fire academy.

S an Jose, California, has one of the best collections of fire apparatus of any American city, including the first little hand pumper bought by the city in the 1850s and this delicious 1937 American LaFrance 750gpm Metropolitan engine that was once the department's pride and joy. Fifty years ago it ran as Engine 1, the flagship of the department. The paint and elaborate gold-leaf are original and, despite being somewhat thinner than when new, show a standard of excellence unequaled today. Lower right, San Jose's first little fire engine, "Old 41," was not quite the very latest technology when it was bought in 1854, but it was the first line of defense for the town until better pumpers came along around the time of the Civil War. San Jose, like many older American cities, has kept many of its obsolete, out of service fire trucks and engines. But few American cities have kept so many and such a rich diversity of old rigs. Besides this vintage hand pumper, San Jose owns two steamers, one of only two Knox-Martin tractors in the country, plus over a dozen other antique rigs.

The firehouses of the downtown area have been stripped and are empty so unassigned truck and engine companies from around the city—and from the neighboring city of Santa Clara—are moved into the stations to take up the slack. Even with the move-ups, the city becomes dangerously exposed to other fires as the incident absorbs resources.

Truck 1, the 85-foot snorkel, applies its master stream to the flames, now finally exposed. In the middle of the wide street out front is Air 6, the rig that recharges the SCBAs worn by the firefighters; all across the fire ground the ringing alarm of the SCBAs sounds as the air runs out. Some captains and firefighters ignore the ringing and stay inside, eat the smoke; there are times when there is no real option.

The Division C commander calls the incident commander with a hopeful report: THERE'S A GOOD INTERIOR ATTACK GOING—HOWEVER THE CEILING IS STARTING TO COME DOWN! THE CREWS INSIDE ARE TAKING A BEATING. THEY ARE BEING SUCCESSFUL BUT I NEED ONE MORE HAND LINE BACK HERE!

The IC calls back: COPY. I'VE GOT ENGINE 2 COMING IN FROM SANTA

CLARA, DROPPING THEIR OWN SUPPLY LINE. I'M GOING TO ASSIGN THEM TO YOU. . .YOU CAN USE THEM FOR INTERIOR ATTACK.

The Truck 1 crew on the roof watch for signs of collapse while they cut with their saws. Those signs come early: portions of the roof start to soften and smolder. When the roof is opened, smoke and flame are released—a bad sign. Inside, the ceiling is full of flame, corroding away the structural integrity of the building. Now, parts of the ceiling begin to fall in on the firefighters. The back portion of the building is doomed. It is time to go.

On the radio, the voice of Division C commander is careful and clear: VACATE, VACATE, VACATE! He calls in at 17:38 on the command channel, ordering everybody out of the doomed single story portion of the building. The dispatcher repeats the order, sending it out to all units in her calm but insistent voice: ALL UNITS INSIDE THE STRUCTURE AT 1401 SANTA CLARA. . . VACATE, VACATE, VACATE! Inside the structure, the captains hear the order and repeat it to the firefighters. You can hear the order on the radio—muffled and distant, with fire noises mixed in as somebody keys his mike in the excitement. "Vacate! Vacate! Vacate!" yell the captains inside. The firefighters in the building throw down their hoses and tools and abandon them to the fire and retreat. Over the radio comes an inadvertent, anonymous response to the vacate order: "Shit!"

Outside, the captains count heads and report to the IC. Everybody gets out, although three members of Engine 8 have been injured—none seriously. The IC calls for a fourth alarm. Then, at forty-two minutes past the hour, four minutes after the vacate order, the roof comes down on the back portion of the building. The IC shifts the offensive attack to the two-story front part of the building. Division A, the first-alarm companies, are assigned a defensive attack to keep the fire from spreading to adjacent structures.

Water supply becomes a problem early. Truck 1 gets a supply line 1,400 feet long, and another truck will use 1,950 feet of hose to reach a hydrant. The San Jose Water Company (the local public utility) is asked to boost pressure, which they do, but even with the supply maxed out the capacity is less than desired. Although all the

engines are on line and pumping 3,500gpm—well over 1 million gallons total—it is still not enough to put the fire out.

Engine 101's captain, John Charcho, is relieved briefly for a quick rest and a shot of lemonade at the Fire Associates rig just outside the fire line. One of the parishioners begs him to get the band instruments out of the still-standing (but now involved) front portion of the structure. The band is supposed to be getting on a plane in a couple of hours to perform somewhere and all their uniforms and instruments are inside. He downs some of the lemonade and wades back into the building, stacking tubas and other instruments outside.

Engine 16 is assigned to interior attack of the upper portion of the two story part of the hall. Capt. Bob King and his two firefighters take a charged 1 3/4-inch line in and head up a broad flight of stairs. They climb into the smoke—and King's helmet bangs into a ceiling hidden in the smoke. Sometime in the past, the staircase had been partially covered; now it is a blind lead. Engine 16 retreats, tries another route and finds a way up to the next floor.

Here's where old fire engines go to die— an anonymous warehouse filled with faded glory. This is only one part of San Jose's marvelous collection—the rigs that aren't currently drapeable. That's a 1914 American LaFrance wooden aerial on the left; the engine for the tractor is being repaired, but everything else is original and works. There's a 1923 "Bulldog" Mack back against the wall, a couple of Seagraves from the same decade, plus some derelict rigs that are beyond redemption and are kept for salvage and sentiment.

This glittering 1925 American LaFrance engine is Firefighter Juan Diaz' pet project. He and other firefighters have spent many hundreds of hours restoring the little rig to its original glory, and it is now a near duplicate of the kind of engine that rolled to many thousands of fires around the country over the years. Fire engines, then and now, are expected to be in first-line service for about twenty years before retirement. But even after all those years of cold starts and quick getaways most rigs still have plenty of life left in them and can continue on almost forever with only occasional attention to the cosmetics.

Outside, Firefighters Juan Diaz and Mike Shaw from Engine 16 are momentarily unemployed. They find a 2 1/2-inch hand line laid in from a hydrant for an engine that has been removed for safety, leaving the hose neglected and available. So Diaz borrows a nozzle from Engine 8 and applies it to the supply line—the direct approach. Although they are at the mercy of fluctuating hydrant pressure, there is enough at the nob to work with, and they apply the stream to the front of the hall.

At this point there is not much to be done except the most basic kind of exterior attack firefighters call "surround and drown." The master streams from each of the engines and from the snorkel are all poured on the flames.

Despite the best efforts of the assembled multitude, the front part of the building becomes involved; at 19:14 another vacate order is issued. The fire promptly engulfs the second floor. The structure is lost.

Out by the command post, by the street and just inside the fire line, the chiefs exercise command. Twenty feet away, on the sidewalk, the parish has gathered to watch their building, a member of the community, die. A tiny old Portuguese lady argues with one of the police officers in broken English; she objects to the loss of the building and its contents. The officer does not have a good answer for her, but he stands there and listens anyway.

The fire ultimately escalates to seven alarms, with control attained about midnight, seven hours after the first call. Fourteen engines responded, including two from an adjacent city; fourteen trucks also responded, plus many specialized rigs.

Although the Five Wounds hall fire was a large one for our city and a devastating experience for the church community, the incident is something that has been happening to cities for thousands of years. In fact, fires like this one happen in one city or another around the country every day. Cities like New York have them almost daily. And, in fact, this particular fire was a mild experience compared to the ones that used to be so common in the United States. Around 1800, "going to the fire" was considered a form of evening entertainment in Boston and New York.

Structural fire emergencies are simply a fact of life and a cause of death, now as in the past. Some of the equipment, tools, and tactics used to fight the fire are new and modern while others are ancient. Fire is a timeless experience in the history of civilization. And, as the Five Wounds hall demonstrates, not even the best efforts of contemporary technology can always beat it.

The business and technology of firefighting is a special one. Firefighters are our last pure heroes. Their big engines and trucks still inspire awe in children and adults who are awed and inspired by nothing else. It is an extremely risky business, with injuries a normal part of the job description and a short life expectancy a statistical fact.

Fire engines and firefighters are both special as a result of the role they play. Both the people and the rigs are tough, reliable, dependable, and dedicated to the public welfare. All they do, day in and out, is help people. Both are just a phone call away in an emergency. Citizens cherish both.

In San Jose, we have a wonderful collection of old rigs that goes back over a hundred years. We have beautiful old Seagraves and Macks and American LaFrance pumpers from the 1930s that have served for decades and have been replaced but not discarded. We kept one of our hand pumpers (nearly 200 years old now), our old steamers, and the hose reels that once were pulled by teams of men through muddy streets. We still have a 1914 aerial ladder rig with its chain drive tractor and huge tiller wheel. One of our steamers pumped for a hundred hours at the San Francisco earthquake and fire—and even though it is just a piece of old, worn-out, obsolete machinery, that little bit of legend has been preserved along with the pumper. Even the hard, cold machines of firefighting are heroes to a lot of us. Fire engines do that to people.

We still have a few firefighters from the 1930s, too—and they're old, worn-out, and obsolete as well. But they haven't been forgotten or discarded, either. Men who operated the 1937 American LaFrance at a hundred fires, when it was the hottest thing in town, still visit with the men and women in the department today. Their pictures are on the station walls showing them young and energetic at major fires fifty years ago.

Although my city is on the West Coast, it is in most respects a contemporary of Eastern commu-

"Vacate! Vacate! Vacate!" yell the captains inside. The firefighters in the building throw down their hoses and tools and abandon them to the fire and retreat.

25

nities—especially in the business of fire suppression—and is representative of the American experience of firefighting. San Jose was founded in 1777; the first formal fire company was organized in 1854, about the same time similar companies were forming everywhere. The city still owns the little hand pumper, built in New York about 1800, that served the city of San Jose back then. It was operated by the same kind of rowdy volunteers that served all American cities at the time. It came out here along with a couple of early pioneers who had been members of a New York volunteer company; they bought the little engine for $1,600 and it went west with the wagons.

The city still has some of the old hose reels that came along after the Civil War, too—examples of the latest high-technology of the time. It wasn't long after the first steam-powered pumper was demonstrated in the East that San Jose bought one; the city still owns two. Any other kind of municipal equipment would have been melted down decades ago and forgotten, but not our fire engines. And I think it is like that all over the country, with old dusty relics tucked into warehouses or parked out in the corner of a maintenance facility—equipment that is just too wonderful to allow to die.

So there really isn't anything special about our department. If anything, it is just one of a hundred medium-large-sized cities around the country that started out as farm towns with little volunteer departments and grew, along with the times. We went through all the same phases as towns like Schenectady, Milwaukee, Rockford, or Memphis at about the same time, from buckets to hand pumps to steamers to automotive rigs. There is something universal about the business of fighting fire. Firefighters then and now belong to a universal community; it was—and still is—common for a firefighter visiting from another town or country to drop by the local fire station, get acquainted and have a cup of coffee, assured of a friendly welcome anywhere, anytime.

San Jose's 1931 Mack blasts along under the control of a appreciative firefighter. Fire engines, unlike almost any other kind of vehicle or in any other nation, inspire a kind of affection and delight that has become a part of our national legend and lore. Part of that may have something to do with the work of these trucks, but some of it may have to do with the love and skill and attention to detail of their builders, too, who have always emphasized quality in design and construction for these very special kinds of vehicles. Although the 1937 "Alfie" is considered by many to be the quintessential American fire engine (and by almost everybody to be one of the most beautiful trucks ever built) solid, trouble-free Macks like this one are often adored by firefighters because of their tenacious reliability.

Chapter 2

Fire Fundamentals

Although the big, beautiful fire trucks and engines have become an important part of American folk legend and lore and much admired, few people understand what these rigs do and how they were developed. To really appreciate and understand fire engines, one needs to understand the basics of firefighting. The following section is a condensed version of what fire academies teach their students.

The business of a fire company is saving lives and preserving property—in that order. For hundreds of years, that has been the basic set of missions for firefighters and their rigs. The mission statement applies to the little Vermont volunteer company with an old pickup truck and the biggest engine company in the Fire Department of New York. Each department's mission is tied to a particular community with its own particular needs. Those needs dictate what kind of equipment, people, and training the department requires. The apparatus is designed and built to meet those requirements. That is the function that determines the form of truck or engine a particular community buys.

No single rig can do everything on the fire ground very well—although many (called "combinations") make the attempt. Fire suppression is a team sport, with many specialists. And the first thing to learn about specialists is that there are two kinds of basic rigs, *trucks* and *engines*. A fire *truck* carries ladders and tools and sometimes hose, but it does not pump water. A fire *engine* is a pumper,

the vehicle that puts the water on the fire. Trucks and engines are normally staffed by from two to five people and are each usually called a *company*. Each rig and the crew assigned to it become a company and the person in charge of the company is its *captain*.

Engine 8 is typical of a small urban company. Its captain, engineer, and firefighters are part of Battalion 1, the downtown area. Other rigs—an 85-foot snorkel, several aerial ladder rigs, engines and water tenders, the air rig, rescue trucks, and light units—are stationed elsewhere around the city at strategically located firehouses intended to provide overlapping coverage and mutual support in major emergencies.

Upon arriving at a fire ground, the engine company makes the initial size-up, and its captain assumes command. He automatically becomes the IC and is in charge of the fire ground until relieved. This first-in company has two responsibilities: to save lives and search for the seat of the fire.

The size-up is the evaluation phase—the captain must determine the seriousness of a situation. Are people's lives threatened? Are multiple structures involved? What response is required, with how many people and what equipment? When dispatchers hear, "Flames showing. . .structure fully involved," they know the situation is serious.

When a first-in engine company rolls up to a fully involved structure fire, the captain will order the driver/engineer to drop a 5-inch supply line

A classic, 1960s, collectable fire engine. This beautiful 1963 American LaFrance pumper was purchased right out of service for $2,000 in 1992, complete with much of its original equipment and in virtually perfect condition. There are a few problems with the ownership of such a rig; one is putting up with all the requests for free rides from neighbors; another is finding a place to park it, and when you take it in to be detailed, they charge $500 for washing and waxing it.

This is really what a fire engine is all about— to enable these firefighters to actually fight a serious fire. In a few seconds this hose team will execute an interior attack on a fully-involved structure fire.

hose at a hydrant on the way in (unless there happens to be a hydrant conveniently close, which is unusual, but possible). The driver/engineer will use the front-mount connection, with its short section of 5-inch preconnect. One of the firefighters will hop off the rig, pull the coupling and first length of the supply line, and wrap it around the hydrant while the rig drives up to the structure. The supply line will come peeling out of the hose bed; if the driver goes too fast, the big metal cou-

plings will dent or damage the engine as they fly out the back. The captain will order the engineer to park in the best tactical position available.

Although every fire is different, there are standard operating procedures (SOPs) for what happens next; typically, the captain will order one of the firefighters to pull the line. The engine will usually have 200 feet of 1 1/2-inch or 1 3/4-inch preconnect hose with a big, variable pattern nozzle attached. The hose is flaked (coiled neatly so that it will play out smoothly, without tangling) into its own hose bed by the pump panel with two big, exposed loops sticking out. Firefighters take the nozzle over their left shoulder, put their right arm through the two exposed loops, turn away from the engine taking the load of 200 feet of hose on their back, and, if they do it right, walk up to the front door and drop off the hose nice and neat. Even while dry and empty, the hose is a 100-pound load so it must be extended now, while it is as light as it gets. Like so many other things in this business, firefighters practice extending the hose; some have turned it into a minor art form. Even though firefighters call the tangle of hose lines on the fire ground "spaghetti," the line should be laid out in a way that it can be pulled into the structure with a minimum of effort.

If there are lives to save, important property to rescue, or if the building can be saved, a company mounts an interior attack. It is dangerous to enter a burning building—it is weak and will collapse sooner or later. Modern equipment and training allow firefighters to go deeper and work longer under worse conditions than in the past. Some important features of the contemporary fire engine that did not exist on earlier models are the provision for SCBA bottles, specialized rescue tools, and nozzles for different kinds of attack.

As one firefighter pulls the line, the captain and the other firefighter follow, bringing forcible-

Here's one of New York's many big Seagrave fire trucks, an aerial ladder rig. Note the distinction between a fire truck and a fire engine—a conventional fire truck lacks pump, water tank, and hose. Combinations of the two have been common for nearly a century. *Piet Halberstadt*

Trucks and engines are usually found working as a team on the fire ground—the engine crew executing interior attack, the truckies searching the interior, making rescues, laddering the building and (as a merry little band of house-wreckers are doing here) cutting holes in the roof to ventilate the structure. The old "Alfie" in the foreground wears its original paint but thousands of wipe-downs and washes have pretty well polished off the lettering that once proclaimed its identity. Both the engine and truck have enclosed cabs, once (and in some departments, still) considered to be sissy stuff. When the truck crew drops in for a call they knock on the door with a 5-pound sledge hammer or any of a wide variety of forcible-entry tools. It is not usually a subtle business, and it can be quite dangerous. All trucks and most engines carry many such specialized tools and this load, rather than improved pump capacity, is responsible for making today's fire apparatus larger than those of forty or fifty years ago.

entry tools. The engineer waits at the panel, ready to feed water. If firefighters are lucky and the fire has ventilated, they will hear the fire "talk" with rumbling and roaring noises. The sounds of breaking glass are music to a fire captain's ears because that means the interior probably will not be full of flammable gasses waiting for the breath of oxygen that comes when the front door opens. When the gasses meet the oxygen, they explode, blasting the firefighters across the street. What firefighters do not want to see as they approach a building are little puffs of smoke coming from under doors and around the eves. Those are signs that the building is sealed and full of superheated combustible gasses.

"Charge the line!" is the captain's order back to the rig, delivered in the loudest, adrenaline-filled voice available, reinforced with a hand signal—both arms upraised like the touchdown signal in football. The engineer opens the discharge gate on the panel and the hose surges to life, hissing and squirming as the water rushes in. The hose will writhe and often come up off the ground, and the firefighter on the nob will feel the weight of the water.

Firefighters on the nozzle bleed the hose line—they crack the nozzle a little to let the compressed air in the line out, otherwise they could just blow compressed air on the flames as they begin the attack. The firefighters set the nozzle to a

Modern fire engines typically carry several different kinds of hose, each with its own assets and liabilities. There is about a quarter mile of red 5-inch in this hose bed, enough to effectively bring the output of a fire hydrant directly to the fire with virtually no pressure drop. The 5-inch, the 1 3/4-inch, and the 1 1/2-inch here are all carefully flaked to drop off the rig without tangles or hitches.

33

Big fires generate big messes. When several engines drop lines on a fire the resulting maze is called "spaghetti." One of the problems to be dealt with after the fire is out is finding your lines and picking them back up—long, hard work particularly after a hard few hours on the nob. This fire was caused by a young arsonist who decided to put a match to the huge rolls of paper in the background.

30-degree pattern and crack the valve—it burps and belches for a moment then water comes out in a nice, strong pattern. Once water gushes out, firefighters call, "Ready!" to the captain.

"Hit it!" the captain commands. The door is forced—either with an ax, halligan, or kick (which looks great on TV but is not recommended)—and the firefighters move into the building, nozzle valve open; they swing the nob in a rapid, clockwise rotary sweeping motion, dragging the heavy hose behind. The room may be full of smoke and steam (firefighters navigate by touch, banging off furniture and walls) or if the room is well ventilated, they can see where they are going. Firefighters communicate with the people behind them by

screaming through the SCBA mask. If the room is involved, they will scuttle along the floor, on their knees or side, keeping low and out of the worst of the heat, while the fire runs the ceiling overhead.

The firefighters advance on the fire, knocking it down as they move ahead toward the worst of the flames. It can take a lot of water to extinguish the fire, and it all turns to steam until it is cooled. A firefighter's life depends on the engineer at the pump panel and the water supply being pumped to the nozzle. If the fire is bad enough and the building is not breathing well, the heat can drive firefighters right down to the floor. It can be so dark that the firefighters might as well close their eyes—the effect is the same. The room can also be perfectly clear, except that there is a smoke ceiling about 4 feet off the floor below which everything looks normal, above which is solid smoke. This smoke is actually a flammable gas, and can—probably will—"flash" on firefighters when a fresh supply of oxygen hits it from somewhere; then they'll be able to see real well.

As in war, the best attack is usually fast and aggressive. The booster tank, and the chemical tank that preceded it, were designed for this kind of quick knockdown. But both have the dangerous limitation of short duration supplies; even a modern rig has only enough water aboard for a three- to five-minute attack.

The noise inside an involved building can be deafening. The fire roars, steam hisses, and the structure creaks and crashes. Glass breaks and the "truckies" on the roof work their K-12 saws and chain saws full blast. All too soon, the SCBA alarms start ringing, signaling that only five minutes of air are left in the air tanks. Radios squawk incomprehensibly, and firefighters yell at each other through the face masks of the SCBA.

Once the team finds the seat of the fire, the firefighters attack it with circular sweeps of the nozzle. The water turns to steam, expands, and makes the atmosphere a kind of hot, foggy hell. The fire will, as they say, "darken down." The team has "made the save."

While everybody stays alert for rekindles, firefighters begin the salvage and take-up phase. Even though the fire is out, the material in the building can still be very hot and close to the point of ignition. And there is a terrible mess to clean up before the captain is released and the company takes up its hoses, stows all the gear back in the lockers, and returns to quarters.

When an engine company responds to a structure fire, a truck company always goes, too. They stay in contact by radio with other responding units—usually through the dispatcher, sometimes directly—to coordinate and control the attack. Meanwhile, the truck company ladders the building. Their job is to get on the roof, break a hole in it for ventilation, and then help with the interior attack. This once, just to be nice, they'll take care of the supply line to the hydrant for Engine 8; usually that's a job they do for themselves but they're a little busier than usual today.

As the Five Wounds fire (described in Chapter 1) shows, getting water inside the hall does not get water on the fire. The truck companies use pike poles to rip out walls and ceilings to reveal the fire; it is filthy, exhausting, dangerous work. The heavy turn-outs and helmet are intended to insulate the firefighter from the radiant heat and falling embers that are inevitable in interior attack.

Getting at the fire often involves taking the building apart, to one degree or another, with saws, poles, and sometimes axes. As a consequence, firefighters were once known as "house-wreckers," because their major control technique involved removing a roof or tearing down a perfectly good building that was likely to contribute to the conflagration's spread. While that is less necessary today, ventilating is still a part of the firefighter's bag of tricks and it still infuriates homeowners as they watch a truck company up on their roof.

Usually a fire will have to be contained or insulated from adjacent structures. Containment is typically done with the big deck gun on top of the fire engine or at the top of the aerial ladder—a high-volume appliance that can shoot a stream of water over 100 feet. This creates a curtain of water that cools much of the radiant heat from the burning building. The deck gun is far too powerful for even a group of firefighters to restrain, typically delivering 1,000gpm or more.

The fire will ultimately come under control and will finally be put out. Then comes the dreary process of overhaul, cleaning up the mess, and finally take-up of the hoses, pike poles, ground ladders, and other equipment that tends to get scattered across the fire ground. The companies return to quarters where firefighters wash and wipe down the rig and prepare it for the next call. The company captain will report to the dispatcher that his company is back in service. With any luck, there will not be another call until the company gets something to eat and perhaps some sleep.

The Fire Engine
Firefighters use two kinds of fire engines—the big city kind, which specializes in traditional engine responsibilities, and the small town or rural version, which does everything associated with

The engineer opens the discharge gate on the panel and the hose surges to life, hissing and squirming as the water rushes in. The hose will writhe and often come up off the ground, and the firefighter on the nob will feel the weight of the water.

engines, trucks, and rescue companies all rolled into one. In a small department, an engine company will do everything, including the ventilation, rescue, extinguishment, salvage, and overhaul. In a large city, the engine company "puts the wet stuff on the red stuff" and not much more.

Ever since the first little hand pumper went to work at the first, long-forgotten fire, the engine's duty has been to put water on fire. Although today's rigs put out more water, are easier to use, and get to the fire faster, the principal is unchanged. And a surprising amount of engine capabilities today are hardly different from those of 1870.

A fire engine provides two basic sources of water to fight a fire. One is out the end of a hose and a nozzle; about the most any firefighter can direct is 350gpm before the force starts waving the nob around—with or without the firefighter still attached. The other is the master stream, a big but normally immobile appliance that is typically a part of the rig. These big deck guns will put out 500gpm or more (sometimes much more) but they are usually only as mobile as the engine itself. The first fire engines, those ancient hand pumpers, had only their gooseneck nozzle and no hose—so firefighters fought the fire with a 100gpm deck gun as well as they could.

Since the hydrants in most cities offer about 70 pounds per square inch (psi) of water pressure and perhaps 2,000gpm flow, who needs a fire engine, anyway? At the Five Wounds hall fire, Firefighter Juan Diaz borrowed a nozzle and applied it directly to the supply lines—he bypassed the engine. But of course it is not always that easy.

The fire engine is designed to supply two or more good sized hoses with water flows of at least 250gpm. That is more complicated than it might sound unless you consider that the pressure in a hose drops dramatically as distance increases. For example, if water at 70psi is pumped in one end of Engine 8's 1,400 feet of 3-inch hose, firefighters will only get a dribble out the other because of the internal friction and expansion of the hose. If they want 70psi and 250gpm at the nozzle, they may need 300psi at the pump, depending on the amount and diameter of hose they have dropped. To get that, they need to have a fire engineer running the pump panel that knows his or her busi-ness, who can calculate the amount of pressure drop for the kind of hose in use, and who can set up the pump to boost the hydrant pressure so it is correct where it counts, out on the nozzle.

Fire engines usually carry two or more heavy, rigid black hoses—even the ancient hand pumpers had them and the ones on new rigs look virtually identical. These are suction lines, normally 5-inch on modern rigs, which firefighters use to draw (draft) standing water from ponds, rivers, bays, or cisterns. The hoses need to be rigid because they would otherwise collapse. While this kind of water supply is seldom used in urban areas, it can still be a critical capability, for example, at a big water-front fire.

While many similarities exist between the mission of a fire engine today and one of a century ago, the new ones carry much more equipment for many more assignments. In fact, the line between engine and truck has become badly smudged, with each often doing some of the other's duties. But as explained earlier, while engines carry some ladders, forcible-entry tools, and similar "truckie" gear, the engine normally has more hose, a much larger water tank, and much more capable pump than found on a truck.

As one long-time volunteer fire engineer and driver says, "There's a lot more to operating a fire truck than just driving it from point A to point B. You need to learn to look, listen, and feel. I look at the gauges, listen to the engine and the pump, put my hands on the hoses at the discharge ports. When that engine is on-scene, the officer and the crew are charging lines and forcing entry, making an interior attack—the engineer has one of the most crucial jobs: getting water to protect the fire-fighters and to put out the fire. When I'm pump-ing, the rig is vibrating and shaking—the RPMs are 'talking' to me. If I'm 'tanking' from the water onboard the engine, and the tank runs dry before the feed line from the hydrant is charged, the engine makes an incredible sound—it screams! You *never* want to have that happen!"

"Truckies"

Rivalry is part of the heritage of the fire ser-vice—between departments, companies, and responsibilities on the fire ground. Members of truck companies sometimes publicly disdain the

Fully involved. Once a fire gets going with a good supply of fuel and air it does not take long for something like this to occur, even with modern construc-tion methods and materials. The walls of this house have burned off in less than a minute and in just a few minutes more the whole thing will collapse.

A 1943 American LaFrance advertisement that plays on the fact that firefighters are often the sons, and now daughters, of firefighters.

work of the engine companies and call them wimps; if you want to be a real firefighter, they say, join a truck company where the real work gets done.

"A truck company does so much more than an engine company," says one truckie, smirking immodestly. "Truckies are known to be extremely large, with extremely large egos. In the old days, as well as today, the ground ladders on the truck can weigh over five or six hundred pounds. They take five or six strong people to move and erect. To perform rescue, to chop holes in roofs requires tremendous endurance, strength, and agility. Everybody knows that being a truckie is the most important job in the fire service!" Engine crews, on the other hand, maintain that truckies are really just highly-paid vandals, but that is just part of the traditional rivalry.

The primary mission of the fire truck is rescue, ventilation, salvage, and laddering. Trucks often carry a pump, some hose, and a small booster tank of water. They can even put out fires, but normally have other assignments. The truck company is expected to support the engine company, both working as a smoothly integrated team. The truckies usually follow the engine onto the fire ground and tie their pump into the hydrant with a short length of 5-inch that normally comes off the front of the engine. Then the truckies go off with barely suppressed glee to wreck the building with saws and pike poles and axes.

To complete a job quickly and efficiently, a truck carries pry bars, ladders of different sizes, circular and chain saws, and pike poles of several configurations. The truck also carries life-saving equipment such as Stokes litters, back boards, and sometimes sophisticated devices such as defibrillators. For automobile wrecks (a common call), there are often specialized Hurst tools (jaws of life) and air-jacks for raising vehicles away from trapped victims, along with sets of blocks for supporting the wreckage while the rescue effort is underway.

Despite the different equipment, many trucks look an awful lot like engines, because the trucks carry pump panels, booster tanks, and even a good supply of hose. This is particularly true of modern snorkel rigs and aerial ladders, which are set up for a primary mission that involves putting water on fire. Of course, both of these rigs are also set up for rescue assignments and their crews are normally assigned truck duties—so that makes them trucks.

That He, Too, May Choose His Vocation

Men of AMERICAN-LaFRANCE are working 'round the clock for a victory that will insure his freedom of choice.

THAT look of admiration for his father's fire engine is a heritage from grandfather's day. For grandfather too, looked with pride on his AMERICAN-LAFRANCE steamer. And should young Johnny follow in the footsteps of his father and grandfather, you can be sure that his AMERICAN-LAFRANCE equipment will still be the world's finest.

The AMERICAN-LAFRANCE * trade mark, for generations of fire fighters has been the symbol of dependable, up-to-the-minute fire protection. It is your symbol of security and protection, on the finest fire apparatus money can buy —built by the world's oldest and largest manufacturers of fire apparatus and fire fighting equipment.

* LAFRANCE-FOAMITE in Canada

AMERICAN-LAFRANCE-FOAMITE CORPORATION
ELMIRA, NEW YORK, U.S.A.
LAFRANCE FIRE ENGINE AND FOAMITE LIMITED
TORONTO 9, ONTARIO, CANADA

They call it "take up," the dirty business of collecting the hose, draining the water and putting all the equipment back on the rig before heading back to the barn. Then the rig will get washed and chamoised— and the firefighters will too.

Chapter 3

Vollies and Steam

Firefighting in the Pre-Automotive Era

The role of civic-minded volunteers has been the foundation of American firefighting from the very beginning—and continues today. In fact, the vast majority (about 75 percent) of firefighters in this country and Canada remain volunteers who train and respond to fires on their own time, in very much the same way as 200 years ago. Volunteer companies even operate in major metropolitan areas; New York City is served by several companies of volunteers. Volunteering is a grand tradition, one handed down with many others from the early colonists, with a rich heritage and vital function.

Early American colonists suffered frequently from catastrophic fires. Initially, little could be done about these fires other than attempt to contain them by pulling down adjacent structures. If possible, people would salvage the contents first. These efforts were impromptu and informal in the early years. In 1653, approximately one-third of Boston was destroyed in a single fire. By 1679, the city had ordered a little hand fire engine, a used single-action pumper probably built by Thomas Lotte in 1743. The little rig needed twelve men to operate it and their efforts produced an effect about the same as that of a garden hose, except that the stream was intermittent. And since this was before the hose, the stream had to be delivered from wherever the engine was located.

In the days when firefighters hauled pumps themselves from the firehouse to the scene, a chief characteristic of firefighting was that the firefight-

ers were almost always volunteers. At first, volunteers (commonly called "vollies") included just about anybody who happened along. Men, women, and children fought fires. By the time of the Revolutionary War, clubs of civic-minded gentlemen owned and operated firefighting apparatus.

Boston acquired another engine in 1704 and the city's two engines were used against another terrible fire in 1709. But by this time, a rabble of common citizens had taken over the pumps in what must have been a rather chaotic exercise, so by 1714 fire societies had developed in Boston—primarily to salvage building contents rather than extinguish blazes. The fire societies had plenty of opportunity to show their stuff because small fires were frequent at the time.

By 1733, Boston owned seven engines and the technology of firefighting blossomed. About this time, there were six classes of fire engines. They ranged from the smallest, a Number 1 that had a 30 gallon capacity and about a 75 foot range (and cost about £20) to the biggest, Number 6, with a 170 gallon capacity and a 120 foot range (and a price tag of £70 freight on board). These engines were heavy, awkward to move on their small wooden wheels, and were limited to exterior attack only. The pumps were the most primitive single-action type—meaning that they only delivered water on one stroke of the piston. The more capable pumps had suction hose and fittings that permitted them to draft water from creeks, wells, or tubs of water that were in range of the scene.

This early hand pumper was built by James Smith of New York in the early 1800s for a New York firefighting team. It served New York for approximately thirty years before it was shipped to San Francisco in 1850. After a disastrous fire, San Jose's citizens decided they needed a real pumper, so they purchased the rig from nearby San Francisco in 1854. Known as "Old 41," the rig was paraded around town before being delivered to a new fire hall built in its honor. It served the citizens of San Jose for many years before it was replaced. It remains part of San Jose's collection, remarkably preserved more than 100 years after its retirement and after the rigors of more than fifty years of service.

The very first fire "engine" combined dozens or hundreds of buckets like this one with enough people to form a chain connecting the fire scene to a water source. Such a bucket brigade could deliver several hundred gallons of water per minute to a fire but it did an inefficient job of actually using the water—and the "engine" tended to wear out pretty quickly. Nonetheless, households were all required to own such buckets and most were decorated somewhat like this nice one from Philadelphia. The Franklin Fire Society was just one of many private, civic-minded groups of volunteers in the 1830s although by this time the bucket was more likely to pour its water into the tub of a little hand-operated primitive pumper than to slosh it directly on a fire. This one is in a private collection and its first owner, William Love, would probably be astounded to know that his pretty, simple bucket commanded a thousand dollars when it was last sold.

Although New York had a fire company as early as 1658, the company was a patrol rather than a firefighting organization. The Hudson Engine Company, the first real engine company, came along in 1737 operating an English Newsham pumper. The New York Volunteer Fire Department was also formed in 1737 and started out with a home-brewed engine that used metal in its construction and was therefore known as "Old Brass Backs." The idea of the formal volunteer fire company caught on and soon many small organizations, with names like Friendly Union, Hand In Hand, and Heart In Hand, had formed.

In Philadelphia, Benjamin Franklin helped start the Union Fire Company, a select group of thirty gentlemen. Other companies followed the Union model and by 1742 there were twelve companies in the city.

By the time of the Revolutionary War, fire engines and the business of firefighting had both become much more involved. Many manufacturers built the engines, which were becoming more effective. One engine built by the firm of Perkins & Jones could deliver about 300gpm to a range of about 180ft. The engine companies were select, proud societies normally including the most respected and prosperous citizens—the country club of the day.

Volunteer companies were well developed in the East by the early 1800s and were (and still fre-

Old 41

San Jose, despite its provincial location, was quite typical of cities of the mid 1800s. Levi Goodrich and Abe Beatty, two early leading citizens, had been members of a New York City company and had run with "Old 41," a little hand pumper stationed at Delancy and Attorney streets on the Lower East Side. The engine traveled to San Francisco in 1850 while its two former attendants continued on to San Jose. That year, Goodrich and Beatty started the first real fire company in San Jose. It was grandly called Fire Engine Company No. 1 until somebody noticed that there was no actual fire engine in town—or, for that matter, anywhere within 50 miles, nor money to buy

one. The name was changed to Eureka Fire Company No. 1.

But Old 41 was not far away, and in 1854, after a disastrous fire, San Jose found the money to buy it from San Francisco. The little rig had served New York for thirty years or more, at least from 1820, and caused a sensation when it was delivered. The rig was paraded through town, garlanded with flowers, and delivered to the new station that had been built for it. A feast was provided for the leading citizens. The prettiest girl in town, Miss Mary Crane, presented Eureka Company No. 1 with a large, embroidered silk banner from the ladies of the town.

quently are) rather like the Chambers of Commerce or Rotary clubs that enlist the solid business-class people of middle and upper class. With the rapid emigration from east to west that began about 1845, these traditions and the equipment went west with the pioneers.

By about 1850, the volunteer companies in many of the older parts of the country had been transformed from congregations of the social elite to hangouts for the marginally criminal class. The volunteer era developed strong traditions; some of which were dubious. One tradition put great prestige on the company that arrived first at the scene of a fire. The company that arrived first "owned" the water closest to the fire and would rather fight than share. In fact, fights between volunteer companies at fires were common, caused by petty and (to us, at any rate) obscure breaches of honor.

Actually, part of the incentive to achieve "first water" was financial. The insurance companies of the day sometimes provided payment to the fire companies, but only to the first company that arrived and put water on the fire. While this payment was not a factor in every city or at every fire, the practice sometimes turned civic-minded, socially prominent citizens into bands of reprobate ruffians.

About the same time, a series of large immigrations from Europe swelled the population of all American cities. Millions of Irish and German families all competed for a slice of the pie. It was common for communities and institutions to be organized along ethnic lines, in and out of the firehouse. These divisions produced more friction; and the friction generated a lot of heat. It became common for fire companies to be assembled of strictly Irish or German immigrant members, and no one else need apply. A tremendous amount of tension grew between these new companies and the old, established ones of well-heeled, "blue-blooded" citizens of old colonial backgrounds.

Dr. Peter Malloy, director of the Hall of Flame, described the era. "While not all the companies were like that, especially away from the major cities," he explained, "many were like rival gangs that wanted to establish their authority by intimidating other fire companies. The expression 'plug ugly' came from Baltimore about this time, from a policy of some companies to send a brute

to guard the closest fire plug against all comers. But this was just an extension of the politics of the time and the resentment against the Irish, in particular, who were coming into the country at this time. But this was really a big city problem and I haven't seen much evidence that it happened in the smaller towns of the time."

The big city companies of the pre-Civil War era seemed more interested in fighting each other, often, than the fire. But since the big engines of the time required huge numbers of brawny men (brawny women at the time being disallowed),

This little Newsham 4th size hand pumper was built about 1730 and is identical to one purchased by New York in 1731. It delivers about 60gpm, roughly equivalent to the output of two garden hoses. This pumper belongs to the Hall of Flame collection.

43

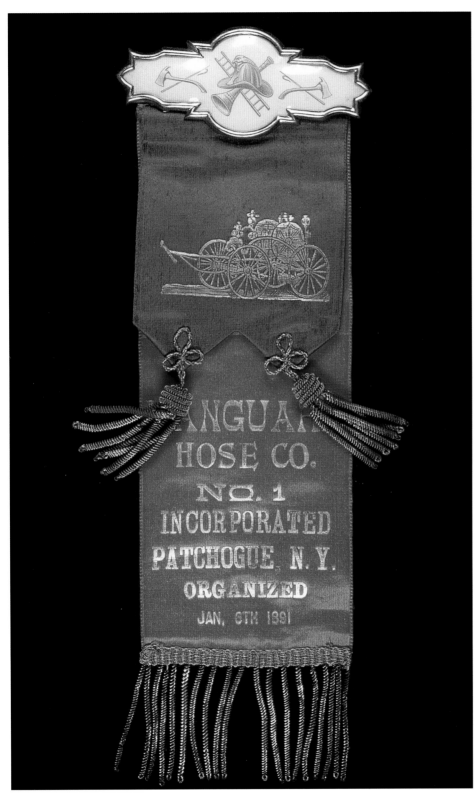

VANGUARD
HOSE CO.
NO. 1
INCORPORATED
PATCHOGUE, N.Y.
ORGANIZED
JAN, 6TH 1891

and since they were all unpaid and consequently independent, there was not much that could be done.

Brawls were common and the biggest became legends. Fire axes, pike poles, water streams, bricks, and almost any other improvised weapon were considered acceptable. Cincinnati, Ohio, was the scene of one of these epic battles in 1851 when a total of thirteen fire companies ignored a burning mill and fought each other instead. The mill burned to the ground—and it was the proverbial last straw for the leading citizens of that fair city who began looking for an alternative to the vollies. That alternative turned out to be a paid fire department, under municipal control, and operating steam pumpers with smaller crews than the big hand engines required. But that was still some years away.

In the meantime, the pattern was the same for American cities everywhere. San Francisco (and even little San Jose) had their brawls and outrages. It got so bad that volunteer companies recruited street thugs and brawlers with no interest in fighting fires at all; these people were expected to help fight other responding companies. The object of the companies became competition with other companies and the preservation of honor. It was considered a great disgrace to have another company beat yours to a fire, and to be passed on the street while your company dragged the engine to the blaze was almost unforgivable.

When relay pumping was done, each company tried to "wash" the next in line by sending them more water than they could handle. A hand engine that had never been washed was called a "virgin" or an "old maid," and was (as elsewhere) scarce. It was also common for one fire company to enlist a member to, through fair means or foul, take control of the available water near a fire before the actual company arrived. This was sometimes done by placing an overturned barrel over a fire plug and stationing an innocent-looking person on it. Other techniques were downright nasty; one New York vollie was notorious for simply cutting off the hose of other companies in order to attach his own. Since this firefighter was reputed to be eight feet tall, wear special shoes with embedded spikes (for fighting), and fight fires fearlessly as well, he usually had his way.

Hunneman hand pumper, 1866. The long black tube is called "hard suction" and is used to pull water from cisterns, creeks, horse troughs, or any other available standing source—exactly as the hard suction on a modern engine does. This rigid hose is carried "squirrel tail" style. This pump delivers 130gpm. This rig served Rockport, Maine, for many years and acquired its name, *Pacific*, as part of a running joke with Rockport's neighboring community of Camden. When Camden named their identical Hunneman pumper *Atlantic*, Rockport (a couple of miles to the west) firefighters snickered a bit and came up with their name. The motto of the Rockport volunteers was and still is Be Early and Cool. This pumper is part of the Hall of Flame collection. Far left, back in the 1890s, as today, firefighting was largely a volunteer business. But back then fire companies often included the cream of society and were often the center of the social life for many communities. Ribbons, pins and other objects were distributed to commemorate these events.

Rumsey hand-drawn Badger pumper, built in 1865. This little rig helped fight the great Chicago fire of 1871 after being shipped by train from its home jurisdiction in Centerville, Wisconsin. The Rumsey represents what is called a "piano-style" engine because of the shape of its tank and pump housing. This pumper is now part of the Hall of Flame collection.

Every proud company seemed to have its own fancy uniforms, and some even had special parade apparatus that never went to fires and was strictly for show. In fact, many vollies participated in the companies to help their social status rather than their fellow citizens. In 1860, one of San Jose's firefighters complained of going to an alarm and getting his glittering boots muddy.

The companies were organized along ethnic and occupational lines and were quite cliquish. The Irish operated San Jose's Empire Company while the Germans worked at the city's Torrent Company. While the companies might (or might not) cooperate to put out a fire, the normal procedure at the end of fighting a fire was to turn the hose on any rival companies present—and perhaps fists as well.

End of an Era

When San Jose decided to take over the business of fire protection from the volunteer companies in 1876, many people objected—and you could hardly blame them, in a way. The vollies had paid large sums out of their own pockets for the apparatus and facilities, and had invested much time and effort to protect the town. Now the city told them that they were not needed any more. While there were riots and mayhem in some cities, as in Cincinnati, San Jose's volunteers went out quietly. As midnight approached on October 31, 1876, the last day of their service, the men of the Empire Engine Company, Torrent Engine, Eureka Hose, and Hook and Ladder No. 1 paraded one last time through the dark, ungrateful streets. At exactly midnight, they deposited their apparatus on the steps of the city hall and went home.

Nobody laments one tradition of the steam era: the practice of using political considerations for the hiring, firing, and assigning of professional firefighters. During the late nineteenth century, newly elected mayors kicked people out of government jobs. A person could have been doing a great job for years, might be close to retirement, and make the mistake of being a Democrat after the election of Republicans to city office—or the other way around. Then that person would be out on his ear, or maybe up on charges for some invented offense. It was a nasty way of doing business, and it prevailed for many years.

Pre-Automotive Technological Developments

Both the social and technical aspects of firefighting evolved rapidly during the 1800s. The Industrial Revolution changed everything for everybody, including firefighters. All kinds of people invented, patented, and marketed new inventions for fighting fires. Fire engines progressed from small to huge hand pumpers, then to steam pumps that transformed the business of fighting fires. Cities started growing up as well as out, streets were paved, telegraph communication became possible, then common, all with implications and opportunities for fire departments to become more effective at saving lives and property.

Hook and Ladder

In the late 1700s, American buildings began getting taller and fires on the roof of three-story structures were becoming more common. In fact, roof fires had always been a basic tactical problem for firefighters. Hooks (for pulling flaming roofs off rafters) and ladders (for climbing to small blazes in chimneys or on roofs) were part of the tools of the trade long before engines. But with the increasing height of urban structures, especially after about 1820, the need for effective aerial attack became acute.

Early ladders were carried on wagons that firefighters pulled by hand. Construction of these wooden ladders became an art and science. Ladder construction itself was the subject of much study and invention. Builders found that a 75-foot ladder was the longest single ladder they could erect; if the ladder was built longer than that, it tended to be too heavy or too weak. The side pieces were normally made of clear fir, a light and strong wood; the cross pieces were typically hickory—extremely strong if not as light as fir. These ladders would reach all the way up to a five-story window, once erected, but were clumsy and awkward to employ. Besides being difficult to move and emplace, the ladders had the unfortunate habit of burning rather easily.

Part of the problem was solved with extension ladders. These came along after the Civil War and solved some of the problems—but added a few of their own. One of the first extension ladders failed during testing, dropping two firefighters to the street far below. Both firefighters were killed, and so were the chances of the new system's adoption.

Detail from a Currier & Ives print. During the 1850s steamers and hand-powered engines often fought alongside each other as portrayed here. At first it was difficult to predict which technology would break down first, but steam gradually displaced the huge volunteer companies and their monster pumpers—along with their monster egos. Left, the gallant firemen from Philadelphia's Hand-In-Hand Fire Company No. 1 pose for the camera outside their station around the time of the Civil War. Most wear parade regalia hats and capes. *Cigna Museum and Art Collection*

47

Parade belts from the late 1800s were part of elaborate and expensive uniforms worn by the "vollies"—volunteer firefighters—on holidays, not to working fires. These belts are from the collections of Zoltan Szucs. Far right, the title page from an 1885 Ahrens-Fox catalog.

The company that arrived first "owned" the water closest to the fire and would rather fight than share. In fact, fights between volunteer companies at fires were common, caused by petty and (to us, at any rate) obscure breaches of honor.

Hose

Although hose was invented in the seventeenth century, it didn't become a practical reality until the close of the eighteenth. The first American fire company dedicated to using hose did not form until 1803 in Philadelphia. Hose really was one of those inventions that revolutionized firefighting. Now firefighters could take the water much closer to the fire, instead of applying it from out in the street. For the first time, effective interior attack was an option.

But it wasn't long before the new tool was used in an innovative way—one still just as important as then—relay pumping. When a fire developed far from a ready source of water, engine companies would set up on line, feeding water from one engine to another until the water could finally be sprayed on the flames. These relays could include over thirty engines and could cover more than a mile.

Hose was (and is) bulky and heavy. The early solution to get it to the fire was to send it along on its own cart, wagon, or reel. The first of these was built in America in 1819 and the idea caught on. Hose reels were typically of two basic sizes, small and light (known as a "spider," because of its spindly construction) and heavy (called "crabs"), carrying about 600 and 1,000 feet of hose respectively.

By 1823, there were nineteen dedicated hose companies in Philadelphia alone. New York ultimately had sixty-three volunteer hose companies to support the fifty-seven engine and eighteen hook and ladder companies.

Getting Up Steam

Of all the wonderful, odd, misunderstood technologies used in firefighting, the steamer takes the prize. Few people today even recognize a steamer as a kind of fire engine. But from the day they were introduced until they finally gave way to gasoline power in the 1930s, steamers transformed the business of the fire ground.

Although forgotten today, except by a few dozen die-hard railroad fanatics, steam technology once was as exciting and liberating as semiconductors and microchips are today. When the bugs were worked out of steam power about 1850, American and European industries were quickly and enthusiastically transformed. Steam power-

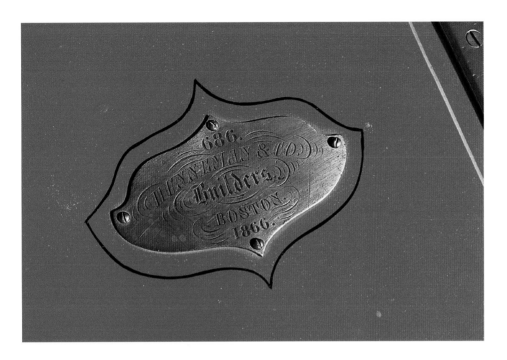

Builder's data plate from the 1866 Hunneman hand pumper.

traditional hand pumpers survived into this century in small jurisdictions with plenty of manpower, steam dominated everywhere else within a few years.

At first the steamers were about as heavy and agile as railroad locomotives—something many steamers closely resembled. Some were self-powered and would creep slowly from firehouse to fire under their own steam. But within a few years, the idea was refined and perfected to the point where a team of three horses could pull one at a gallop through the streets.

In fact, some of the steamers of 1860 were the approximate equal of most pumpers today in terms of pumping capacity and response time. Not only could they put the same amount of water on a fire as a modern engine, the response time for horse-drawn steamers could be as good or better than that of a modern engine company. By the end of the Civil War, in 1865, a top-of-the-line steamer (rated a double extra first) could put out 1,300gpm; 1,250gpm is still considered good and many companies use engines that deliver only 1,000gpm. What's more, some urban companies were set up for extremely fast response and could hitch the horses to the engine, get up steam, and put water on a fire four blocks away within 2 1/2 minutes.

Although expensive to acquire and maintain, steamers were reliable and effective. Before the age of steam, firefighters could contain a fire; with a steamer you could put it out. Philadelphia was one of the last big cities to convert from hand pumpers and vollies—but by the end of the era had purchased 136 steam pumpers over the course of fifty-nine years. That experience was representative of the bigger American cities of the late nineteenth century, although New York's numbers were even larger. But the smaller cities and towns found the steamer to be a good investment, even after motorized systems came along. Philadelphia bought its last in 1914, but kept it working until 1927.

Among the best of the manufacturers was a company called Silsby. Their pumpers were so reliable and exceeded specifications so much that Philadelphia decided to replace its existing fleet with that make. The city bought forty-one Silsbys between 1886 and 1897. But there were dozens of manufacturers competing for business, advertising

plants revolutionized transportation and industry with economical, efficient, reliable power for ships, trains, and industrial production. For the fire service, steam power came at just the right time to solve the problem of the volunteer companies, with their large numbers of rowdies. The large crews were essential to operate the brakes of the huge hand pumpers, but the traditional rivalries and fights between companies—often at the expense of the burning structure—infuriated most citizens, but what could they do?

Until January 1, 1853, citizens could not do much. But then two Cincinnati, Ohio, engineers demonstrated a steam-engine-powered water pump in a competition with the Union Fire Company and their big pumper. Although the Union Fire vollies got first water on the fire with a 200-foot stream, the steam contraption beat them with a 225-foot output. The volunteers pumped away for half an hour, then collapsed. The steamer just kept chugging along, putting out water. The vollies were disgusted—and shortly thereafter they were also replaced by a smaller, paid department operating steam apparatus.

It took awhile to work out the kinks in the technology, but it did not take long for cities around the country to adopt steamers. While the

aggressively. As the population of the country grew rapidly with waves of European immigrants, new cities were developing everywhere, all at once. The opportunity for manufacturers was tremendous.

Some of the manufacturers built steamers as a sideline. Most of these companies just made a few, then went on to something else. Toward the end of the century, several manufacturers formed new conglomerates, one of which was the American Fire Engine Company, soon to become the American LaFrance Company.

Steamers were manufactured right up until 1914 and continued in service until just before World War II. In fact, many were hauled out of storage during the war, including San Jose's 1898 American, a 600gpm unit that was purchased from the city of San Francisco. (This engine pumped for 100 hours continuously during the firestorm of 1906 when the great earthquake broke gas and water mains and most of San Francisco was destroyed.) Although many steamers went to the scrap pile during World War II for their brass and

A Short Course in Steam Pumping

Over the years, many changes have occurred in the business of putting out fires, but the most modern pumpers don't really do the job of "putting the wet stuff on the red stuff" any better than the steamers of a hundred years ago. In fact, the old rigs have capacity to spare, and—what with the trend to recycle—maybe some municipalities will dig their Amoskeags and Silsbys and Ahrens out of the museum and into the line. If so, here is how to get up steam.

Working steamers are kept warm, rather than cold, one way or another. Most have provision for quick water disconnects to keep the boiler water hot at all times. On the grate is kindling, soaked in kerosene, and a supply of coal in small, fist-sized chunks. At the first tap of the alarm bell, a firefighter ignites the kindling and disconnects the water lines while another company member hitches the horses—all of which take mere seconds.

The firefighters left the pumper configured for quick action. Water will be in the boiler; a sight glass will show the level. It is the engineer's job to manage this part of the operation; water level management is crucial while the steamer is in operation. All vents and petcocks for the steam part of the pump must be open at first, until the metal warms. The lubrication system, an oil pump and grease fittings, must be properly configured. The water pump outlet and valves must be open—the outlet on a positive-displacement piston pump like the ones on most steamers cannot be turned off without breaking large chunks of metal.

The suction inlet must be connected to a hydrant or other source of water. Two 1 1/2-inch lines go to the nozzles; each will deliver about 250gpm. If the water supply is clean, it can be used for boiler feed, otherwise a clean source has to be used.

Within minutes the pressure gauge will come up to working pressure—55 pounds or so. Then the engineer opens the valve to let the steam into the pump. Large volumes of water vent at first from the drains. When that stops, the engineer closes the valves and petcocks and the pump begins to work, slowly at first as the pressure builds.

By the time the firefighters get on scene, the boiler fire roars. (There are two kinds of boilers, fire-in-tube and water-in-tube; the first is used by locomotives, the second by fire apparatus because it works much faster—a lot of fire is applied to a little water.) Fire management is the full-time business of the fireman. The boiler fire has to be kept stoked with coal and not allowed to get too low or too high. Once the pump starts working, the exhaust steam vents up through the smokestack, providing a tremendous draft for the fire. This suction can actually lift the fire right off the grate, and it is the engineer's job to adjust the draft control to keep this from getting out of control.

About forty little items need to be checked and adjusted to get the pumper in action and, while they hardly ever blow up, it is quite possible to bend things badly if the fireman forgets something on the checklist. One New York engine company, in 1910, was so well rehearsed that they bolted out the door, were on-scene four blocks away, and had water on the fire in 2 minutes 35 seconds.

51

iron, others served as reserve engines and actually fought fires.

"The steamer grew out of the industrial revolution and the ethnic unrest of the period," Dr. Peter Malloy observes. "The big problem with the steamer was that it only worked in an area with a decent water supply. If you didn't have a decent supply, the steam pumper could use up all the available water—from a well, for example—in thirty seconds. Towns like New York and Philadelphia had state-of-the-art water systems, starting in about 1845 for New York." Smaller towns installed cisterns under the streets, and some (like San Francisco) still have them, and a fire might require hose runs of hundreds of feet from the cistern to the fire ground.

Two problems with steamers were never really solved. The first problem was that they took awhile to get up enough pressure to operate the pump. The other was that they were heavy for the roads of the time, which were usually covered with mud, and the mud added even more weight. Of the two, the first was the most difficult to solve.

Many ways were invented to speed the process of getting up steam. One involved a connection between the boiler and the steam heating system for the firehouse. This kept hot water in the boiler at all times and shortened considerably the process of getting up pressure. Another system used a small pilot light under the engine but forward of the firebox. As the rig moved out of the station, the pilot light ignited the fuel under the boiler. Naturally, kindling and coal were always prepared in the firebox, ready to set alight. Some companies left a match with a nickel near it for the first boy to arrive and start the fire at the sound of the alarm bell.

There is something quite endearing about a steamer in action. Ken Bechthold, captain of the 1874 Clapp & Jones of Woodland, California, says, "It is a *privilege* to operate such a piece of machin-

Cleveland, Ohio's, Engine Company 26 prepares for a run in this posed but charming photograph from March 1912. The engineer is lighting off the boiler fires while the driver reaches for the harness release that will drop the collars onto the waiting horses. One of the crew is already hooking up. The steamer is probably a Mansfield Machine Works 3rd size pumper with a capacity of 550gpm that was purchased by the city in 1901 and assigned to Engine 26. Photographs like this one were fairly common at the time, although the illumination source— flash powder made from magnesium—was notorious for starting fires as well as recording them. *Louis Baus photo/John McCown collection*

52

ery." Watching a steamer in action is a privilege, too. It chuffs along musically, black coal smoke pouring from the stack; sunlight—or the glitter from a burning building—flashes on the polished brass and bright chrome. The Clapp & Jones uses an unusual horizontal pump—most are vertical.

Woodland's pumper has been in city service ever since the day it was bought new and brought across the plains, well over a hundred years ago. Ulysses S. Grant was president and the transcontinental railroad had just been completed. In 1873, Woodland had experienced a fire that wiped out the town, a common but annoying experience at the time. The traditional bucket brigade and hand pumper hardly slowed the blaze, so the small com-munity decided to scrape up the $10,000 for the Clapp & Jones. Another fire came along not too long after and—unfortunately—the steamer was not as effective as the city fathers and mothers had intended. So they dug even deeper and bought another engine and the Clapp & Jones became Engine 2.

Sounding the Alarm

The steamer was just one product of the Industrial Revolution that radically transformed the fire service during the last half of the nineteenth century. Another nearly as important development was the invention of the electrical fire alarm.

The Phillipsborn-Kirby department store at 326-328 Euclid Avenue, Cleveland, Ohio, goes up in flames, November 15, 1908. The horses for the steamers have been led away to shelter, but the team for the ladder rig on the right are still in harness; they've been covered with large blankets as protection against the chill. *Louis Baus photo/John McCown collection*

BUILT BY THE

Detail from a Silsby poster from the 1860s. Silsby engineered some of the finest, most technically elegant steam pumping engines ever built.

One of the problems for the fire service before this invention was the lack of a good way to alert the fire department of fires. By the time somebody had run to the local station with the news, the blaze had an opportunity to get a building fully involved. With the advent of the telegraph and telephone, however, a faster system was possible . . . but it took awhile to figure out how to use this new technology.

One early fire alarm was the Gamewell Fire Alarm box. These boxes were essentially simplified automatic telegraphs wired directly to the fire station. At first, most cities had few of them widely spaced. In San Jose, these were considered far too valuable to be available to just anyone. Consequently, they were installed in houses, the residents of which were supposed to activate the alarm in case of fire in their neighborhood.

Unfortunately, one of the houses selected in our town was indeed a "house"—of ill repute, a certified brothel. It didn't take the girls who worked there long to learn that they could get a whole truckload of firefighters to come at one time just by pressing the button. While this was effec-

tive, and probably a lot of fun for all concerned, the captain of the Empire Engine Company was highly annoyed, possibly because he wasn't on scene. It wasn't long before the alarm boxes were placed on telephone poles.

The Trouble with Horses

Even after they accepted steamers, the vollies resisted using horses for several reasons. One was that horses cost money all the time, whether they were taking the firefighters to a fire or kicking back in the station. Somebody had to feed, water, and tidy up after them; in communities that only had a fire every month or so, that seemed like a large investment for a little payoff. But the steamers were heavy. They took the fun out of dashing off to the fire, mostly because firefighters did not exactly dash when they were pulling 10,000 pounds (some were as heavy as 16,000 pounds) of iron through the mud—they plodded, and it was not pretty.

Horses went to work on the fire ground about 1870 and were, for many years, the delight of the firefighters and the folks living nearby. The animals were treated as pets, given treats, and taught tricks. Before the turn of the century, it was common for folks to drop by the firehouse to visit the horses. The animals had many virtues, once trained, and were the source of much pride and pleasure. As part of a proper team, fire horses were enthusiastic members of the company that could function with a minimum of guidance from the driver. When an alarm sounded, the horses moved into position in front of the engine, on their own. As the station doors opened, the team charged out onto the street, maintaining perfect step. It was a sight to see.

The real trouble with horses is and was that they can't be turned off. They stand there, in quarters, day in and day out, turning horse fuel into horse exhaust. The fuel economy, as a result, was terrible. A standard, one-horsepower fire horse, in a two-bit, dirt-simple little town like San Jose was in 1880, got about 100 bales of hay to the working mile. Of course, they were good company and loved a fire as much as the firefighters, but they still generated an awful lot of exhaust.

When steamers appear in parades these days, they are often pulled by heavy draft horses, but that was never the case a hundred years ago. Fire horses had to be fast as well as strong, agile, and (to the extent possible for a horse) smart. Three were usually used to pull a steamer that weighed in at about 4,000 pounds. They were usually of medium size and perfectly matched in appearance if possible. Sometimes they were specially bred for the job.

Portland, Oregon, used Percherons and Morgans to produce a hybrid with the power to pull the heavy engine through muddy streets and the speed and agility to get to the fire sometime before it burned out on its own. The resulting animals became special pets of their Portland neighborhoods. People dropped by the fire station every day with carrots and sugar lumps for the horses. They were all known by name and were given special treatment that conventional draft horses did not get.

Steam fire engines like this American Metropolitan from the 1880s could deliver about the same volume of water on a fire as most modern rigs of today.

This American steamer will pump 1,000gpm for as long as the fuel and boiler water hold out and could sometimes get on-scene and into action faster than the automotive rigs that started displacing them.

Although the Union Fire vollies got first water on the fire with a 200-foot stream, the steam contraption beat them with a 225-foot output. The volunteers pumped away for half an hour, then collapsed. The steamer just kept chugging along, putting out water. The vollies were disgusted—and shortly thereafter they were also replaced by a smaller, paid department operating steam apparatus.

Retired lieutenant Pat Madigan is a member of a four-generation firefighter family that goes back to before the turn of the century. His father joined the Portland Fire Department in 1911 and Madigan, who is now in his eighties, remembers the horses with affection. "I remember being the only kid in Portland allowed to ride on the steamer behind the horses," Madigan recalls. "The firemen would pick me up and put me on the seat beside

them—I was the envy of every kid in Portland!"

San Jose received its first horses in 1871, five years after the first Silsby steamer arrived (at the astronomical cost of $6,215) in 1867. At first, the husky Irish lads had towed the 4,000 pound monster around by hand, but at last they decided this was folly and did what the other departments were doing around the country, enlisting teams and teamsters for the task.

Another delivery photograph from around the turn of the century, this one of another bit of technological innovation. Below the driver's seat are two big copper "chemical" tanks that could go into action almost immediately.

E ngine 28 arrives on scene sometime around the turn of the century. Three-horse hitches were the norm for fire engines at the time. Horses were selected for speed and stamina over sheer strength. *Jim Hart collection.* Right, the Clapp & Jones company built some excellent steamers with a variety of designs. This one uses a large, dual-action horizontal pump and is still owned by Woodland, California, which bought it in 1874 and had it shipped across the plains and mountains. It takes about a hundred pounds of coal to get it fired up, but then the engine chuffs along in beautiful tempo that varies slightly with the rate of water flow.

The change from large companies of volunteers with their hand-powered pumpers to small companies of paid professionals with steam apparatus was one of the major transitions in the fire service. The first paid firefighters across the country received about $70 per month—for a 24-hour-a-day job. Captains were paid $75, engineers $80, drivers $100 for themselves and the horses. Some only got one day off a month. Nonetheless, some firefighters managed to marry and a few produced children . . . although the kids had a curious tendency to closely resemble the milkman.

Horses had to be selected and trained for the job, a process that took many months and a skillful trainer. But the result was something firefighters and citizens all adored. Through the early years of horse power, a useful system evolved. The animals were stabled near the rigs at the rear of the station, with their harnesses suspended by ropes from the ceiling. At the first sound of the alarm, a firefighter helped the horses move into position in front of the pumper or ladder. At a tug of the reins, the harnesses would automatically lower and the animals would be ready to go in seconds. The horses, trembling with excitement, would strain at the bit, anxious to go. The steamer, with its boiler already hot, would not need long to get up pres-

Cleveland, January 15, 1912. The temperature is 8 degrees, and a fire is gutting the Browning & King clothing store. The third alarm has been called and the horses have been led off to shelter while the long, cold business of fighting the fire is accomplished by Engine Company 28 and their 1907 American LaFrance Metropolitan 900gpm 1st-size steamer. *Louis Baus photo/John McCown collection.* Below, this photograph records an aerial ladder rig from the horse-drawn days, the paint and varnish scarcely dry. Included aboard are axes, lanterns and buckets, plus a seat for the "tillerman" at the rear, a relatively new feature at the time this rig was built.

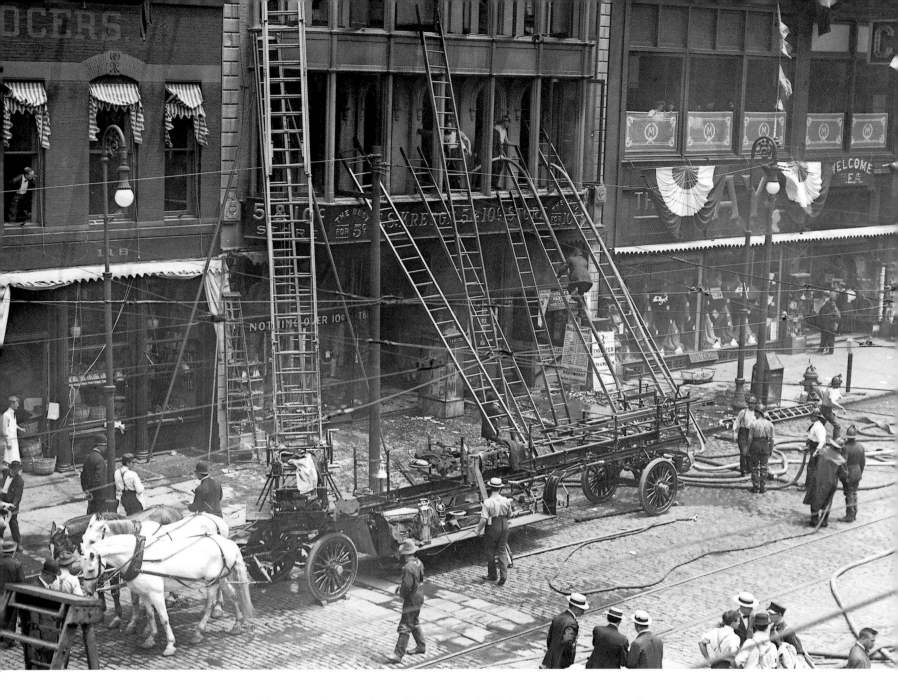

sure. If the boiler fire was laid and lit properly, there would be a head of steam sufficient to pump on arrival at almost any call.

San Jose's Empire Engine Company was timed in the 1880s at thirteen seconds for the process of getting the horses hitched and the pumper out the door; six minutes from the alarm, it was putting water on a fire—about the same response as today's engine companies, and better than some. This speed was the result of regular

drills that became SOP for departments around the country. Horses needed to be exercised and trained, anyway, and the firefighters were getting paid, so by drilling, the standard of performance was elevated and tested.

Although firefighters of that era did not have to worry about the same kind of hazardous materials that challenge today's men and women, they had their own set of special fire problems. One was fires in manure piles; the other was the occasional

Fireworks were once responsible for many major fires, including this one at the S. S. Kresge Store at 2025 Ontario Street, Cleveland, Ohio on July 3, 1908. A child threw a lit sparkler into a table of fireworks, with deadly results. *Louis Baus photo/John McCown collection*

Technological innovations, like this cast-iron hydrant, rapidly changed the business of fire suppression during the last half of the nineteenth century.

San Jose's Empire Engine Company was timed in the 1880s at thirteen seconds for the process of getting the horses hitched and the pumper out the door; six minutes from the alarm, it was putting water on a fire— about the same response as today's engine companies, and better than some.

Chapman Valve Manufacturing Co.,

Manufacturers of

GATE

FIRE

HYDRANTS

—WITH—

Babbitt Metal

SEATS.

These Hydrants afford a clear Water Way of the full Diameter of the Pipe.

The Drip is a positive drip, opened and closed by the Gate, and is closed the moment the Gate commences to open.

No Waste of Water at the Drip. No Water-Hammer on the Pipe in closing.

ALL HYDRANTS GUARANTEED.

Office & Salesrooms,

77 KILBY STREET, - BOSTON.

privy fire in the days before indoor plumbing. Both of these incidents had a tendency to take the fun out of a call. As electricity was introduced and gas lighting became widespread, the firefighters had a full set of challenges from emerging technologies. These electrical fires were in addition to the ones caused by arson, careless smoking, kerosene lamps, flimsy wooden construction, and the use of coal and wood fires for cooking. Fires were common and frequently disastrous.

San Jose's worst fire happened on July 2, 1892, when a fourteen-year-old boy flipped a lighted cigarette into a fireworks vendors' wares. The display was impressive. When the sparklers, roman candles, fire crackers, and rockets were expended, more than forty downtown buildings were afire—including Empire Engine Company No. 1. The battle to contain the fire would have to be fought by the other fire companies and Empire Engine Company, along with much of the downtown, was destroyed.

Driving a team of horses hauling a massive, top-heavy steam pumper galloping through the muddy streets of a town required tremendous skill and a fair amount of courage. But the horses seemed to thrill at the occasion and it was not a problem to get them going at a dead run. Getting other traffic out of the way was the problem, and crashes were not uncommon with other teams, streetcars, and occasional inanimate objects. But everybody cheered as the firefighting team went by, and many ran behind in their wake to the grand event.

The firefighters grew to adore their animals and considered them part of the company. But as automotive technology was perfected in the late 1890s and early 1900s, it was obviously time to consider a change. The new gasoline-powered vehicles were loud, unreliable, and slow—but the exhaust was less of a problem and they did not consume fuel constantly. The economies were obvious, the technology was improving, and it was at last time for the horses to retire.

Many went reluctantly. They were sold for use on delivery wagons, mostly, a common form of transport in some American cities (particularly Baltimore, Maryland) until quite recently. These animals were notorious, in their new employment, for responding to alarms with the same old enthusiasm and would frequently show up at fires right behind the motorized apparatus. Milk wagons in particular were frequently hijacked by reprobate fire horses.

The Last Run

In 1924, Seattle, Washington, sold its horses as Engine No. 24 became the last company in that town to convert to automotive power. The day before the turnover, an announcement was printed in the newspaper; the next day the streets were lined with people. The last run would be to a box alarm, but instead of a fire there would be dignitaries waiting. Waiting, too, were dozens of Seattle citizens who wanted to say good-bye; they clutched apples, carrots, and sugar lumps, the last treats for the faithful horses. The box was pulled and the horses made the last run. After kind and nostalgic words, the teams were driven off to their new homes. Back in the station, a new Engine 24 was already in service.

Pat Madigan was there. "Horses were like human beings to the firemen, they loved them just like people. The firemen cried on the day they left—they were losing their friends! They were only sold to people who would treat them well, put them out to pasture, and treat them like human beings, never to somebody who'd use them to plow or anything like that."

When San Jose's last teams were sold off in 1915, Empire Engine No. 1's driver, Tim Sullivan, took them for one last run. Then he remarked about the new gasoline powered rigs. "They're good machines, all right," Sullivan said, "but you can't love them, and they can't love you." The chief's team was the last to go, and when they went he said of them, "You cannot drive them into a ditch, and if you leave them alone they will never strike a tree, a pole, or other obstruction. They will get you to the fire."

That was much more than you could say for the newfangled rigs around 1907. They were, from a users' point of view, without visible virtue—slow, unreliable, hard to get started, difficult to drive. It was rare that firefighters arrived at a fire, in the first years of their use, in anywhere near the time horses would have made. Many firefighters chose to retire with the horses rather than make the conversion to the new technology. It was a trauma all the way around.

In 1910, New York discovered that it cost only $85 a year for gas and oil for one of the new automotive chemical wagons. A team of horses for the same kind of rig would have cost $660.

This Clapp & Jones steamer is still owned by the Woodland, California, Fire Department, which bought it new.

The San Jose chief's anxiety about the loss of his trusted horses was well founded. His successor, Chief Richard Brown, received a nice big Stutz touring car in 1910. He drove it to a state meeting of fire chiefs and on the way back lost control of the machine and died under the big Stutz.

But some of the firefighters began to take a shine to the new rigs and the chiefs certainly found things to like about the new apparatus. In 1910, New York discovered that it cost only $85 a year for gas and oil for one of the new automotive chemical wagons. A team of horses for the same kind of rig would have cost $660.

The transition from horse to automotive technology was done gradually and sometimes resulted in long periods of training and equipping. The journal for our little Chemical Company Number 2 shows that the horses went out of service on March 23, 1915, and that the motor truck was not in service in San Jose until October 16 of that same year.

Tim Sullivan stayed on and learned to drive the new apparatus, but he never fully adjusted to the automotive age. The story says that for as long as Sullivan drove for the department, he continued to bring his rig to a halt the way he had always done—by firmly yelling, "Whoa!"

The Fire Department of New York got rid of their last team of horses in 1922—there was one last dash out the door, past the assembled, admiring multitude, never to return. Detroit's Engine Company 37 was that city's last to convert. When they went on their last run, 50,000 people lined the streets to watch the beautiful sight for the last time. The end of yet another firefighting era.

Engine Company 14 rumbles through Cleveland's Public Square, a two-horse team pulling what appears to be a 1899-vintage 1st-size Thomas Manning 800gpm steamer. A canvas skirt protects the firebox from drafts while a Homberg hat does the same for the engineer. *Louis Baus photo/John McCown collection*

65

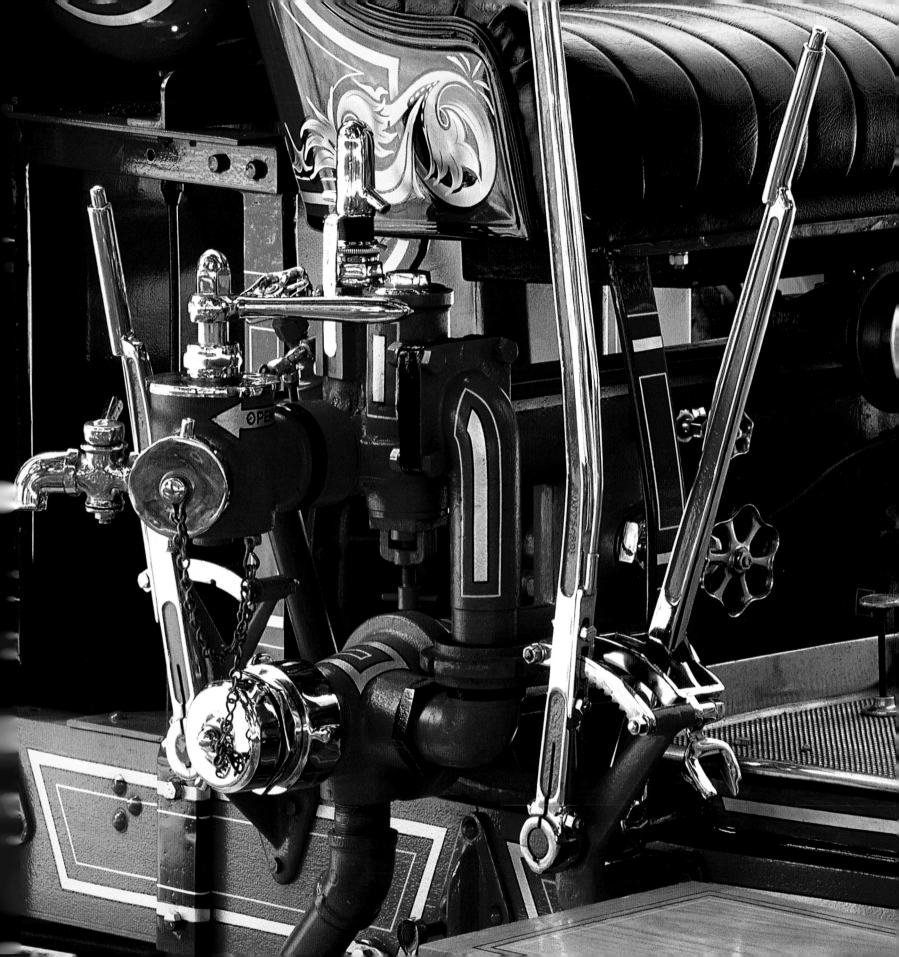

Chapter 4

The Beginning of the Automotive Age

Firefighting has a history of periodic revolutionary changes rather than a gradual evolution of techniques and technologies. It began with buckets and ladders, progressed to hand pumpers, then to steam and finally to automotive power. Much of the business remains in antiquity. Some of the newest rigs, which do not actually pump more water than the steamers of 1870, have equipment aboard that is centuries old in design. Firefighters wear the same leather helmet from 120 years ago, organize their efforts the same way—and some even grow the same kind of mustaches. It is that kind of business, reeking of history and heritage.

Crank it Up

About the time firefighters had worked the kinks out of steam pumpers and horse-drawn transportation, along came the "infernal" combustion engine, much to the distress of just about everybody. But the politicians liked the idea of the possible economies that would result from getting rid of the horses—and their fuel bills.

Actually, the gasoline-fueled, four-stroke-cycle, water-cooled powerplant was only one contender in the battle for supremacy in the automotive technology marketplace around the turn of the century. The first self-propelled fire apparatus of the automotive era was an electric chief's car that San Francisco purchased in 1901; New York bought a similar vehicle shortly thereafter.

Steam power was used successfully in some cars and had been propelling fire apparatus since the middle of the nineteenth century; it was still very much under consideration at the turn of the twentieth. In fact, Vancouver, British Columbia, bought a mammoth 16,000-pound self-propelled steamer in 1908—and it kept on ticking until about 1943 (and went to the scrap yard ten years later). This big engine, built (appropriately) by the Amoskeag division of the Manchester Locomotive Works, delivered 1,200gpm but only had a 12mph top speed; a "code three" response was not an option.

Electrical power was another alternative for the big fire engines and trucks as well as the little chief's cars and occasional hose wagons. It worked well for some cars and it found its way into a few fire rigs. Battery powered vehicles started instantly, were quiet, and could be powerful. A number of chief's cars and light trucks were tried with motors instead of engines but had the pesky habit of running out of juice on the way back from the fire—if not on the way to the scene. One hybrid used a gasoline engine to power a generator that then fed electrical power to motors at the wheels—a complicated and inefficient system, but one that solved the pesky problem of drive train design.

Gasoline power had its own set of vices and virtues. While the tiny engines used to power light buggies around town seemed odd and interesting at first, they were hardly powerful enough to drive a heavy truck. One early application of gasoline

A perfectly restored American LaFrance combination fire engine from the 1920s, attired in a rich, subtle, and accurate dark red.

Waterous hand- or horse-drawn pumper, 1906. The Waterous Pump Company of St. Paul, Minnesota, built the first American gasoline-powered pump, the first version appearing in 1898. This one was built in 1918 for the town of Plainfield, Wisconsin. This pumper is part of the Hall of Flame collection.

power came from the Howe Fire Apparatus Company, a horse-drawn rig with a little gasoline engine driving a water pump.

The First Gasoline-Powered Fire Engine

Most books say that the first gasoline-powered fire engine came along in 1906, but this is not really true. The first gasoline fire engine was an excellent, highly successful little combination built in 1904. The reason it is commonly overlooked now is that it was constructed in England by the Merryweather Company for the Fitchley Fire Brigade. American apparatus builders were

certainly aware of and interested in this development. The Merryweather fire engine used a 30hp engine for both propulsion and, through a nifty coupling system, to power the 250gpm pump. This rig carried a 60-gallon chemical tank, 180 feet of hose, a telescopic ladder, fire extinguishers, and a length of hard suction hose. The Fitchley engine, despite its English pedigree, deserves credit everywhere as the first of the breed; it has survived the years and today resides in the London Science Museum.

There is a bit of dispute about just who built the very first American gasoline-powered fire engine, but it was probably from the shops of a firm that is still in the fire business today, the Waterous Engine Works in St. Paul, Minnesota. That first engine was an ugly little thing, all boxy and awkward in spite of a little pinstriping and red paint, that rolled out the door in 1906 and into service for the town of Wayne, Pennsylvania. The vehicle used two gasoline engines, one for propulsion, the other for pumping. Compared to the Merryweather engine, this little rig was something of a weakling.

Waterous engineers must have done some brainstorming because they came up with a better idea the next year—one gasoline engine for propulsion and pumping both! "Wow!" said the folks at the fire department in Alameda, California, "we gotta have one of those!" The engine arrived with a cute little length of preconnected hard suction hose but not much else; the hose had to go to the fire in a separate rig. This is probably the first combination of truck and engine that was equivalent to the modern solution to the problem.

Waterous revealed their new design in 1907, which turned into a busy year in the apparatus industry. American LaFrance delivered a gasoline-powered chemical wagon and the Webb Motor Fire Apparatus Company sold Tulsa, Oklahoma, a Thomas touring car that had been heavily customized with a pump and ladders. The folks who were still selling and servicing steam fire apparatus began getting indigestion.

The smart ones had been studying the problem for years, trying to adapt to the difficulties of the new technology. Seagrave began a design and test program two years earlier, in 1905, but was not going to rush into the market. After some aborted attempts at developing a suitable powerplant, the company developed and built two gasoline engines, a four-cylinder and a six-cylinder version. One made an epic, historic journey from the plant in Columbus, Ohio, to the distant town of Chillicothe—55 miles over the primitive dirt roads of the era.

But by 1907 Seagrave was ready and sold a trio of rigs to Vancouver, British Columbia. American LaFrance was still in test mode, and would not become active in the motorized market for another few years, but they did sell a few rigs that were modifications of other builders' chassis. The next

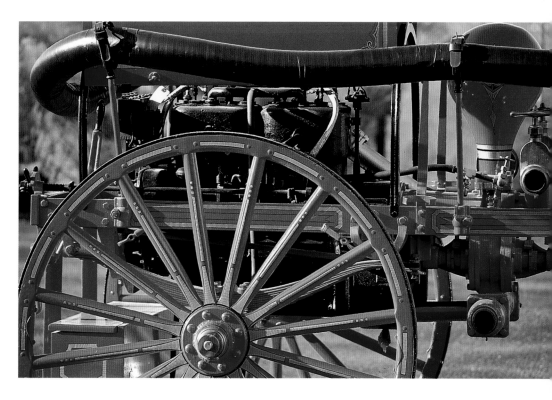

year, they offered a prototype to the Chicago Fire Department who tested—and rejected—the rig.

While steamers were still being manufactured, the companies' sales staffs were beginning to encounter a good bit of buyer resistance to the old product line. Even so, the legendary name of Ahrens (previously incorporated into the American Fire Engine Company conglomeration) was joined to that of Charles H. Fox to form the Ahrens-Fox Fire Engine Company with the intention to build steamers.

It would be awhile before Ahrens-Fox caught on, but the other builders were busy. Seagrave delivered an aerial ladder rig with an automotive tractor in 1909, but it wasn't the first of that particular breed—an International(predecessor to Mack) beat them to that punch. Another company, Monhagen Hose Company, sold the first rig with pump, water tank, and hose—the very first "triple combination."

Within a few more years, the dominance of the gasoline engine was fully established. American LaFrance finally entered the market in August, 1910. Their first official rig, Register Number 1, went to Lenox, Massachusetts, and is still there,

The Waterous was powered with a Waukesha four-cylinder engine attached to a 350gpm pump. It was started with a hand crank and used pump water for cooling. The Waterous represented a huge technological advance for some departments. Those big wooden wheels were perfect for the rutted dirt roads of the time. Rigs like this stayed in service for many years. This vehicle is part of the Hall of Flame collection.

San Jose's, 1914 Knox-Martin tractor is one of two known to exist. It, and the 1898 steamer to which it is attached, both fought the great San Francisco fire of 1906. Far right, the Knox-Martin's driver position is designed to be similar to that of a horse-drawn rig.

eight decades later, in perfect shape. Electrical power was still viable and Ahrens-Fox offered a chief's car of this type while Seagrave sold a massive 85-foot aerial ladder with four-wheel electrical drive. In 1913, American LaFrance sold the Fire Department of New York twenty-five aerial ladder rigs with a combination gasoline-electrical propulsion system. The gasoline engine drove a generator that in turn powered motors attached to each wheel.

The anemic performances of the rigs of just a few years earlier were far surpassed by 1912 when

1,000gpm gasoline-powered engines became common. Seagrave offered 600gpm to 1,000gpm pumpers and, for the first time, a pressure regulator. The Fire Department of New York ordered twenty-five American LaFrance tractors and Birmingham, Alabama, bought eighteen Seagrave rigs—the conversion from horse-drawn steamers to automotive power was, as firefighters say, "en route, code three."

The first gasoline-powered vehicles were a real challenge to the firefighters. San Jose's department started the transition in 1912 when the city rejected all new requests for additional horses. A pair of early trucks were bought. When run through the same trials required of the older horse-drawn rigs, the new apparatus failed miserably—but were retained anyway because of promised economies.

"I actually saw horse-drawn rigs pass up a truck company going to a fire!" says retired Seattle lieutenant Pat Madigan. "Those were the days! The guys today will never experience the thrills I had back in those days!"

San Jose bought a Knox-Martin three-wheeled tractor for their 1898 American steamer in 1914. Both the Knox-Martin and the pumper had served the city of San Francisco, although not together. The combination was christened Engine 1 and turned out to be durable in a limited way. The rig served the city for several decades and survives today. The pumper is rated at 500gpm but will put out 750gpm. The odd little tractor was (and is) difficult to start, unlike the horses, although it certainly shivers when called on to go someplace. There is only one other known to exist in the country, and it resides in the Smithsonian Institution in Washington, D.C.

The San Jose chief's anxiety about the loss of his trusted horses was well founded. His successor, Chief Richard Brown, received a nice big Stutz touring car in 1910. He drove it to a state meeting of fire chiefs and on the way back lost control of the machine and died under the big Stutz. His last run was a slow one, behind horses, a "special call" that turned out the entire fire and police departments of the city, all draped in black.

Another common occurrence was the problems many of the old drivers had of converting to the new vehicles. Well into the 1920s, apparatus builders sent delivery engineers along with new rigs to instruct the driver. The manual that came with American LaFrance rigs seventy years ago advised:

"It is well to confine the course of driving instruction to one or two men rather than attempt the education of every member of the company.

"Two prime requisites for a driving position are the ability to act and think quickly; taking a large pumping car or aerial ladder truck through traffic at the speeds that are sometimes necessary is quite different than handling a Ford. . . .Watch the skillful operator; regardless of the emergency, every

The Knox-Martin used one technology that lasted for many years—chain drive.

movement is not only made quickly but easily and noiselessly; no effort is exhibited at all and the machine is always under complete control."

Most drivers learned to steer well enough, and after some practice they learned to work the gears. Even the Knox-Martin was designed, as much as possible, to simulate the experience of driving a team of horses; the steering column was placed where the reins would have been, for example, and the rig was designed to turn in the same radius. The firefighters of the day, who had learned to be careful around horses, had to learn to be careful with the fire "cars." It was quite easy to have an arm broken or be otherwise injured. Automotive power had pretty well displaced the horse by about 1910 in most major cities; in minor ones (like San Jose) it was just a matter of time. But it would be

another few years before the gasoline engine would be effectively coupled to a pump that came close to the efficiency of the steamer. In the meantime, though, handsome chemical wagons, ladder trucks, hose wagons, and water towers arrived on-scene attached to popping, gasoline-powered vehicles.

The chemical wagon uses big copper tanks of water, typically about 50 gallons, to which a chemical has been added to make an alkali solution. Within the tanks, in a glass reservoir, is a smaller quantity of concentrated acid. When the acid is dumped into the alkali solution a chemical reaction occurs that produces large quantities of carbon dioxide gas that quickly pressurizes the tank. Within a very short time there is enough pressure to squirt the solution out to a considerable distance. It's a simple, reliable, economical system

Nearly ninety years old and still running fairly strong, this early example of gas-powered fire apparatus is still lovingly preserved by the department it served for many years. The suspension and steering systems were advanced enough for their day, and the solid rubber wheels had many advocates into the 1930s, but the technology of automotive systems developed rapidly just after the Knox-Martin entered service, and it was quickly superseded by better designs.

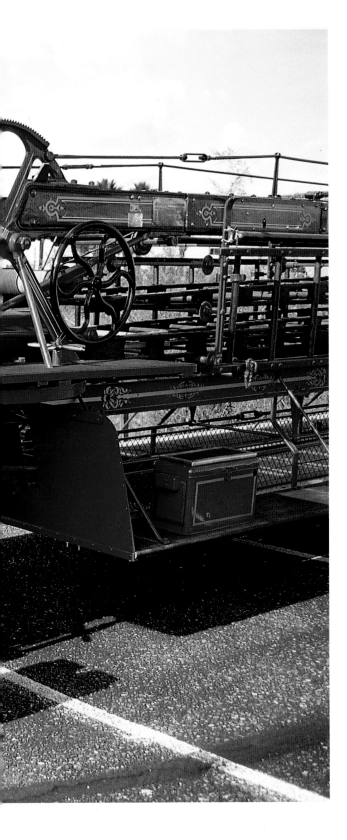

Another early example of early automotive technology, and a more successful one than the Knox-Martin, is this American LaFrance Type 31 aerial ladder from 1919, previously owned by the town of Danville, Illinois, where it served until about 1960. The rig used a two-wheel tractor that was extremely popular around the time of the First World War. The ladder itself was a revolutionary system, patented in 1903, that used huge springs to help raise the ladder to the vertical. This vehicle is part of the Hall of Flame collection.

that was popular into the 1930s, when automotive power, pumps, and booster tanks finally took over.

Getting Started

The automotive system of propulsion may have offered some economies but it sure took some getting used to. No longer did the horses automatically dash off with the engine—in fact, the engine just sat there, cold and silent, unless the engineer did everything just right. As late as 1926, Seagrave provided extremely detailed instructions for starting and driving the rig, written with the presumption that the driver had never operated a motor vehicle before. The manual describes three methods of starting the engine, including one with the old hand crank.

"Starting the Engine by Hand with Magneto Ignition:

"Pull out the compression-release handle on the dash.

"Retard the spark lever to the mid-position on the quadrant.

"Pull the steering-mast controls up to provide a richer mixture for starting.

These drawings from an American LaFrance catalog show two configurations of the Type 31 two-wheel, front-drive tractors. One version was designed specifically for towing steam fire engines and the other was built for towing aerial hook and ladder trucks. The Type 31 was equipped with either a 75hp four-cylinder engine or a 105hp six-cylinder engine. The Type 31 rolled on solid dual tires.

Type 31

Two Wheel Front Drive Second Size Steam Fire Engine

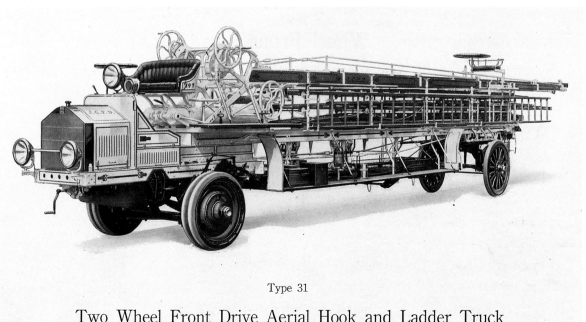

Type 31

Two Wheel Front Drive Aerial Hook and Ladder Truck

Chevrolet chassis were popular with some of the smaller apparatus manufacturers that did not want to build their own chassis. Left, San Jose's beautifully restored 1925 American LaFrance triple combination pumper is typical of 1920s rigs—light, fast, reliable, and perfectly suited for the small town roads of its day.

top, and to stand well away from the crank when turning it. That is because the crank can easily recoil forcefully with bone-breaking force. Broken thumbs and arms were once fairly common, and it was also common for the crank to come around and smack an unwary operator in the chest, knocking him flat.

The compression release handle was an ingenious device connected to the engine's valves. By lifting the valves off their seats through a mechanical linkage, it becomes much easier to turn the engine because the starter wasn't working against its compression. Once the engine started, this control was closed, and the valves functioned normally.

The manual also provided detailed driving instructions. It isn't recorded if anybody propped the book open on the wheel and tried to follow the instructions while driving, but it is recorded that many accidents occurred with the rigs until the drivers got used to them. Following are driving instructions from a 1926 Seagrave operator's manual.

"With the left foot, disengage the clutch by pushing the pedal as far forward as it will go. It is not necessary to put an undue strain on this pedal. The driver can readily tell when it has reached its limit of travel. This operation is commonly known in automobile practice as throwing the clutch 'out.' Hold the clutch 'out' while the gear-shift lever is moved toward the left and then drawn back into first-speed position. Be sure the spark lever is advanced toward the top of the quadrant. Increase the speed of the engine slightly by pushing down on the foot accelerator. Let the clutch in gently by gradually releasing pressure on the foot pedal. The car will now move forward.

"Allow it to gain speed. Again disengage the clutch, but do not push the pedal out as far as the first time; be sure, however, to have it completely released. This position is indicated by the speed of the engine increasing as the load of the car is removed. Remove the right foot from the accelerator pedal so the engine will not race. While the clutch is out, move the gearshift lever forward to neutral, then over to the right, and into second-speed position with a quick movement forward. This shift is most successful if made quickly. (If made slowly and timidly, the gears will probably clash or chatter.) Increase the speed of the engine

Many old fire engines are restored to the same standards as are collectable cars. San Jose's 1925 American LaFrance was restored by the muster team.

"Turn the magneto switch.

"Turn the engine over with a quick upward pull on the starting crank. If it does not start readily, spin it, but never start downward on the starting crank, as great danger is involved.

"Push in the compression release handle as soon as the engine has started."

If you are ever called on to use this technique, remember to also keep your thumb *under* the crank handle, rather than the natural position on

and engage the clutch. Allow the car to attain the speed of about ten miles an hour. . . ."

No wonder they were sorry to see the horses leave.

Christie Tractor

One of the few reliable, powerful engines available to run fire apparatus in the toddler phase of the motorized era was a rather odd-looking machine called the Christie tractor. This was a power unit with only two wheels and a transversely mounted four-cylinder engine in between; if it wasn't firmly attached to something big and heavy, it would flop over on its nose. This unit was attached to many steamers and previously horse-drawn ladder rigs. It might not always have been quite as fast as the horses, but it proved itself to be reliable and sufficiently powerful, the two essential criteria of the moment.

The tractor was the product of the Christie's Front Drive Auto Company and was produced

A 1924 American LaFrance Type 40 chemical and hose car. This lovely engine has a total of 268 miles on its little odometer—a symptom of the very limited number of fires in the rail yards at San Bernardino, California, where it served the Atcheson, Topeka, and Santa Fe Railroad. Right, the chemical tank's 40-gallon capacity was good for a couple of minutes of continuous action, but if that didn't do the trick, it could be refilled and re-charged in 15 minutes or so if there was a source of water around. This vehicle is part of the Hall of Flame collection.

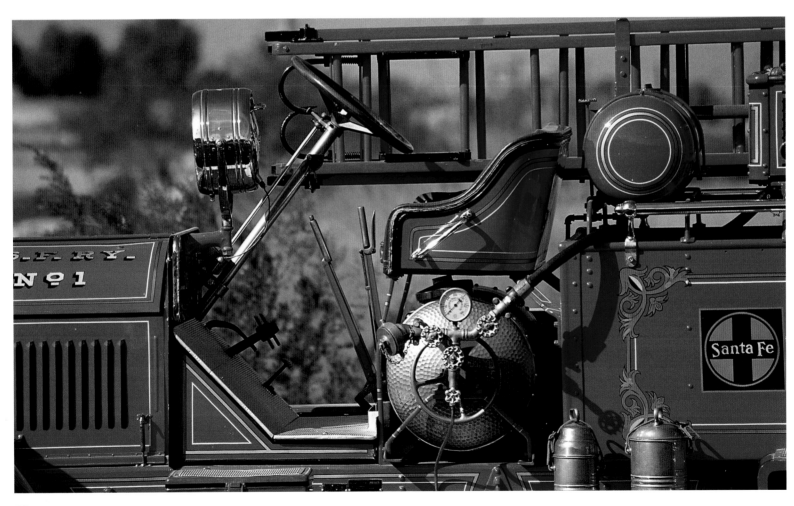

Engine instrumentation on the Type 40 and other engines of the time was Spartan and direct. The Motometer radiator cap was a fixture on thousands of early motorized vehicles, in and out of the fire service; a simple thermometer, visible from the driver's seat, indicated coolant temperature. Top, far left, American LaFrance was still using a right-hand drivers' position in 1924, The driver was responsible for all sorts of chores including setting the spark advance with a control on the steering column, something done mechanically now. Learning to drive was a traumatic experience for many firefighters of the early automotive era and such devices, and the complicated starting sequences of the engines of the time, had many longing for the days when all you needed to do was yell, "Gitty-up!" Bottom, far left, the chemical tank offered firefighters a way of rapidly suppressing smaller fires. The tank's 40-gallon capacity was good for a couple of minutes of continuous action, but if that didn't do the trick it could be refilled and recharged in 15 minutes or so if there was a source of water around. This vehicle is part of the Hall of Flame collection.

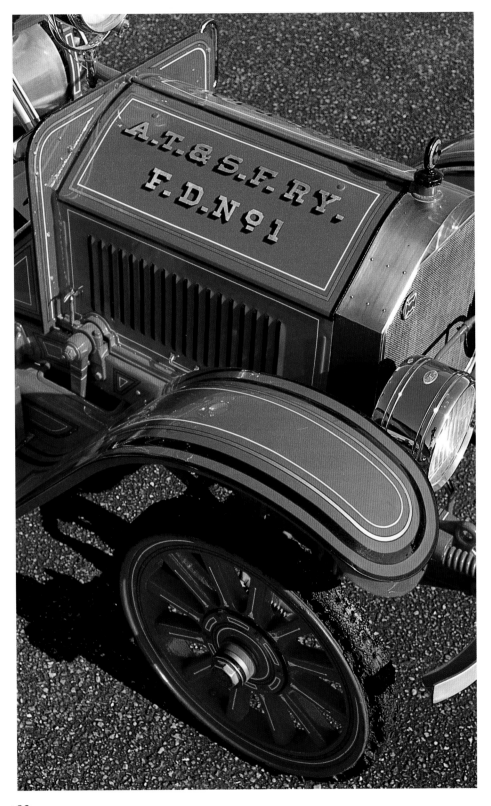

from 1911 to 1919. About 600 were built and bought in large numbers by New York City, which ultimately purchased over 300. It was one of the elements that validated the conversion of the fire service to motorized rigs and demonstrated that fire engines needed a different kind of powerplant and drive train than those used in the smaller, lighter cars of the period.

Post-World War I Developments

Just as firefighters learned to drive, World War I came along. One after another, many firefighters resigned and went off to do another kind of civic duty. The journal of San Jose's Chemical Company No. 2 records their departure—and that one of them pulled a false alarm to celebrate the event. The journal records:

"July 26, 1918: Box 65, 3:18pm; R. Lennon left for camp and pulled this box. J. J. Powers took his place on Chemical 2."

The log records numerous false box alarms during this period, one of the unanticipated blessings of modern technology. It was a common problem for cities everywhere at the time. But entire weeks could go by without a call to a fire, and EMS calls were entirely unknown. Despite efforts to make cities safer against fire, there were still flimsy barns and embers from chimneys, plus a surprising number of arson cases. Chemical 2's journal for November 7, 1920, says, "Special call, 11:15 pm. 500 feet of 2 1/2-inch hose used. Six hours service, Harrison P. Smith 1st and Santa Clara Streets. Total loss."

This period around the years of World War I were exciting for the fire service. Automotive power developed rapidly and new equipment became available. Communities that were still using hand pumpers—and there were some—were going directly into the automotive age without shoveling horse manure or firing a steamer. The Mack AC was offered for the first time in 1915 and quickly gained a reputation for power and reliability in and out of fire departments. It was extremely popular and all sorts of pumps, chemical tanks, and ladders were attached to its chunky frame.

There was much to choose from if you were in the market for a fire engine. Stutz Fire Engine Company, Pierce-Arrow, Packard, White Motor Company, Ahrens-Fox, Buffalo Fire Appliance,

Hale, Waterous, and many other manufacturers offered glittering fire engines with every modern convenience and feature, every choice of model and size, plus just about any color you could ask for (except, it should be noted, lime-yellow).

While many buyers specified various rich shades of red paint for their new acquisitions, there were other beautiful options. Some cities ordered engines in black, dark green, or white. A certain amount of ornamentation was standard and pinstriping and gold-leaf ornamentation was elevated to a high art.

Development of Ladders and Water Towers

Ladders are almost as ancient on the fire ground as leather buckets, but they have lasted much longer and are still part of the equipment of a conventional truck or engine. As American and Canadian cities grew higher in the nineteenth century, firefighters had to find a way to attack fires and rescue people from upper stories. Single piece ladders, lightly constructed of wood and reinforced with outriggers, were effective to about five stories, but no higher. And while early fire engines could

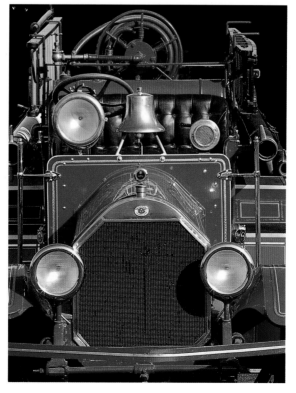

Instrumentation on fire engines of the 1920s tended to be simple and direct. The speedometer almost requires a magnifying glass to read, but the odometer shows a grand total of 2,801 miles for a lifetime of service. The ignition systems dual magnetos have suffered a bit of abuse, though, and one of the switches is broken. But the warning bell control (the rope and wooden pull handle) seem to be as good as ever. Left, a 1921 Seagrave fire engine with a 1,000gpm pump and a huge six-cylinder engine. This rig was typical of the larger engines of the day. Seagrave, while not as well known as American LaFrance, built rugged, reliable apparatus that earned its keep. This one served Phoenix, Arizona, until 1950 and now is part of the Hall of Flame collection. Far left, the 1924 American LaFrance Type 40 still retains a few design details from the horse-power era, including wooden artillery-style wheels with solid rubber tires. This vehicle is part of the Hall of Flame collection.

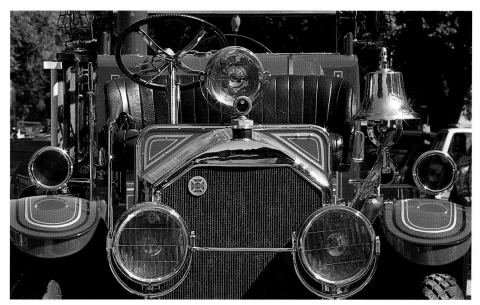

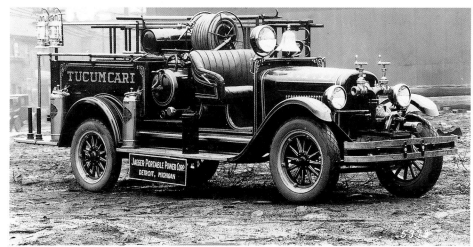

easily throw water to the top of ten-story structures, the stream was not likely to be effective because the angle of the stream, from the street below, had the water washing the ceiling of the upper story rooms rather than penetrating to the seat of the fire.

The water tower was developed to provide effective attack above the reach of ladders. These apparatus were typically steel towers with an extension pipe and nozzle arrangement that could go up to about 70 feet. It was a bit like having a powerful deck gun far up in the air, blasting large volumes of water horizontally into the burning building.

American LaFrance successfully converted to automotive power with the Type 31 and it was

*T*op, here's a 1920s American LaFrance combination. Above, a handsome 1920's conversion, a combination rig built on a Chevrolet light truck chassis. *Hall of Flame collection.* Left, collecting and preserving fire apparatus has become fairly popular and many private individuals have resurrected piles of rusted junk into glittering beauties like this elderly Seagrave.

Right, on early fire trucks the spark is advanced manually by a lever on the sterring wheel. Far right, the shift lever and emergency brake for the 1931 Ahrens-Fox is as massive as the rest of the rig. A hinged lock-out prevents the driver from inadvertently engaging reverse while dashing off to a fire. The artillery-style wheel persisted into the 1930s, finally displaced by all-steel wheels.

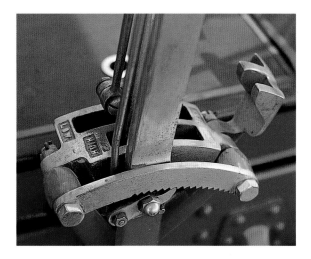

They don't make 'em like this any more—in more ways than one. The accelerator is in the middle with the clutch on the left and brake on the right; try not to get them confused. Note the construction of the pedals—cast steel, designed for eternity. Left, the dual ignition system on this 1931 Ahrens-Fox was a technical marvel in its day, offering superior reliability and relative ease of starting for the massive twelve-cylinder engine. Although called a V–12, the cylinders are actually parallel. Once the powerplant is fired up, the Ahrens-Fox chugs along with a quiet, musical throb that seems to indicate an idle speed of about 20 rpm. It's a bit faster than that, of course, but the sound of this rig, without the interference of a muffler, is a joy.

American LaFrance's spring-assist mechanism for raising aerial ladders was just one of several competing technologies to deal with the problem of fires and rescues in buildings taller than three stories. With this system, large springs raise the ladder to straight up and down, then the operator manually cranks the "fly" section out, and down into position with these large hand wheels. Although seemingly primitive by comparison to modern systems, the length of the ladder was about the same as many of today's. It worked, it saved lives, and at least a few of these systems survive. This one belongs to the Hall of Flame and has been carefully restored by resident craftsman Don Hale.

88

about this time, just prior to World War I, that American LaFrance aggressively began marketing its self-erecting aerial ladders. Prior to this development, getting a long ladder up was a difficult project requiring many firefighters and a fair amount of time. The ladder had poles and ropes attached that the firefighters employed with practiced precision to erect a ladder. Even so, the ladder usually reached no more than 65 feet high. But the Type 31, with huge spring counterweights, made the old systems utterly obsolete; it was a popular piece of apparatus that served with distinction for many years.

The American LaFrance aerial ladder brought revolutionary change to the fire ground. The rig pulled up into position near the structure and the tillerman pulled the pins that allowed the seat and wheel to be swung out of the way while others operated the vehicle's jacks to stabilize it. Two men were stationed on each side of the turntable at the

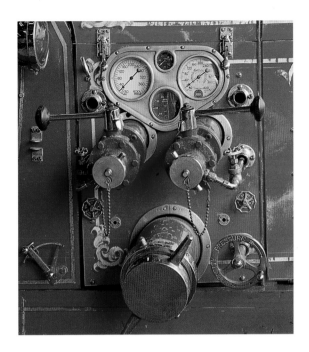

The pump panel from a 1937 American LaFrance "midship" Series 400 Metropolitan pumper. The large connection is the attachment for the feed water from hydrant or, by hard suction. The two smaller ones are the outlets; other controls engage the pump to the engine, adjust output for volume or pressure, and the gauges provide the engineer with data on performance. Below left, a warning plate attached to the 1919 aerial.

SAFETY FIRST

THERE IS NO HARD AND FAST RULE TO COVER FULLY THE USE OF LADDERS. THE COMMON-SENSE JUDGMENT AND EXPERIENCE OF THE MEN HANDLING THEM MUST BE DEPENDED UPON.

THERE ARE SOME WELL-KNOWN PRECAUTIONS WHICH CANNOT BE IGNORED WITHOUT DISASTROUS RESULTS.

BELOW ARE A FEW SUGGESTIONS OF WHAT NOT TO DO:
DON'T USE AERIAL LADDERS FOR "CIRCUS STUNTS."
DON'T ATTACH LADDER PIPE TO FLY SECTION OF AERIAL LADDER.
DON'T OVERLOAD LADDERS.
DON'T SET LADDERS AT IMPROPER ANGLES.
DON'T USE LADDERS FOR SCAFFOLD OR BRIDGE.
DON'T NEGLECT TO INSPECT LADDERS CAREFULLY AFTER USING.
DON'T NEGLECT TO TIGHTEN ALL BOLTS AND NUTS REGULARLY.
DON'T NEGLECT TO CLEAN AND DRY LADDERS CAREFULLY AFTER USING.

SEE INSTRUCTION CARD FURNISHED WITH EACH TRUCK FOR FURTHER INFORMATION.

This 1930 Ahrens-Fox Quad served the community of River Forest, Illinois, into the 1960s and is still in excellent running condition. "Quad" is a fire service expression for an engine with four capabilities: pump, water tank, hose, and ladders; "triples" are far more common. It's huge 1,160 cubic inch engine will still propel the rig to over 50mph or drive the 1,000gpm pump for hours or days at a time. The rig carries a 100-gallon booster tank (as the on-board water supply is called), 200 feet of ground ladders and 1,000 feet of hose. It is a big rig with a big legend. This vehicle is part of the Hall of Flame collection. Previous pages, Ahrens-Fox's big front-mounted piston pumpers were the top of the line during the 1930s—the "Rolls-Royce of fire engines," as Ed Hass likes to say. This novel, gaudy design helped make the line nearly as famous as American LaFrance, despite far smaller numbers of rigs produced. It has also kept many of them alive, and a few are (incredibly) still in service, well over half a century after they were built.

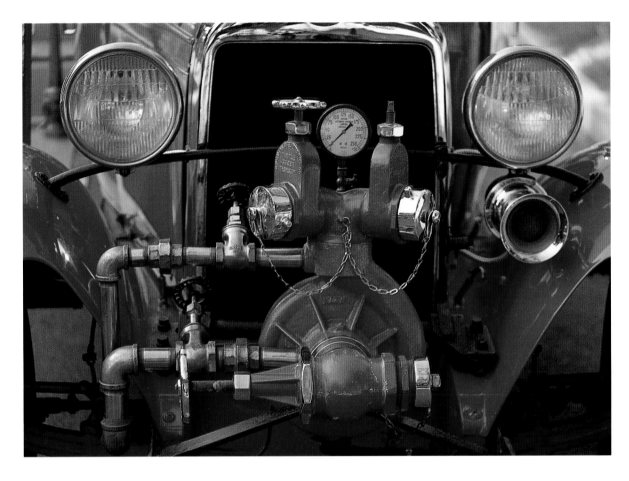

base of the ladder to operate the large geared wheels that positioned the aerial once up. Large springs provided the force to erect the assembly and a pneumatic cylinder regulated the speed of the movement. The operator pulled a safety pin, tripped the lever that released the spring, and the ladder started to rise. A brake lever allowed control of the speed of the process. Once the ladder was fully vertical, it could be moved down and rotated into position with the large hand-driven gear wheels.

The huge ladder rigs normally had straight frames of incredible length to support the ladder length of up to 85ft. Both the front and back wheels were steerable, with a tillerman seated astern with a massive wheel to control the end of the rig.

While the wooden ladders were (and still are) elegant and lovely, they had major flaws and, in the longer lengths, were dangerous. Although such builders as Seagrave and Pirsch & Sons

developed wooden ladder design to a minor art form, the properties of hickory and Douglas fir essentially limited the practical length of a ladder to about 65 feet if it was to be put up by hand, and about 85 feet if it had a mechanical hoist. Even then, wooden ladders were subject to decay, hidden cracks, and often failed under the stresses of the fire ground.

Reports of ladders breaking while firefighters carried down victims from upper stories were fairly common around the turn of the century. The combined weight of a hefty firefighter in sopping turn-outs with a distressed damsel (or more likely a dowager) across his shoulders would apply perhaps 350 to 450 pounds to a section of wood that was already stressed by the support of its own weight—plus the weight of several other firefighters. Under such conditions, the ladder could and often did fail, dumping rescuer, victim, and supporting cast to the concrete below.

Quincy Township, Michigan, bought this beautiful little chemical car sometime shortly after World War I, and it might very well have replaced a hand pumper similar to those in use in the 1700s, many of which were still in service at this time. The rig was built by Prospect, based on a 1 1/2-ton Chevrolet chassis. The chemical tanks are the AC type, holding 35 gallons each and feeding dual hose systems. *Hall of Flame collection.* Left, gold leaf was once common on fire rigs and is almost impossible to duplicate today, so many of the more decorated rigs, such as this 1937 American LaFrance Metropolitan, are preserved rather than restored.

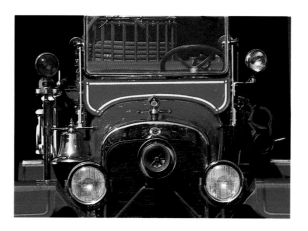

Downers Grove, Illinois, bought this big Seagrave in 1927. Seagrave pioneered the use of the centrifugal pump, now the standard for the world's fire services. Centrifugals are high-speed pumps, well matched to the speed of the internal combustion engine.

As the technology associated with gasoline engines and their drive trains improved during the early years of the century, metal ladders became an option for ladder-rig designers. Steel, and later aluminum, offered superior strength for the same length ladder but with a severe weight penalty that delayed its use until mechanical systems were developed to manage the placement of the systems on the fire scene. It takes the power of automotive engines to operate the mechanical and hydraulic systems that exptend and position these long, heavy ladders with the speed and precision required on the fire ground.

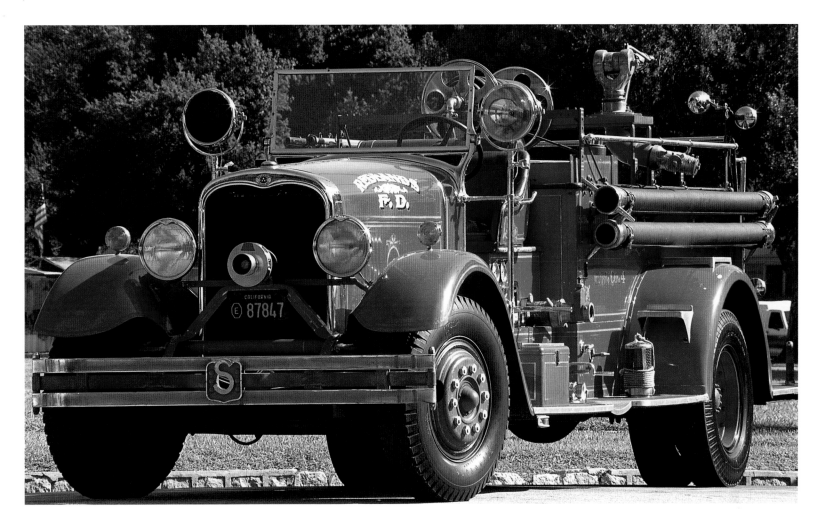

Early aerial ladders used springs and, later, compressed air to erect the ladders from a fixed turntable on the truck frame. A system of gears and cranks, powered by strong firefighters, moved the ladder into position. Then, in 1931 Seagrave offered the first fully powered aerial with hydraulic elevation; extension and rotation were powered by the 160hp engine through a gear train.

While some claim the first motorized ladder was bought by the Fire Department of New York in 1912 from the Webb Motor Company, others point out that actually, it was Seagrave that sold a Mack AC-drawn 75 foot ladder to Vancouver, British Columbia, three years earlier, in 1909. Seagrave—true to its heritage—sold the first all-steel ladder in 1935 and many others shortly thereafter. Pirsch introduced the first all-aluminum

ladder in 1936 and continued to specialize in this corner of the apparatus market.

Ladders today have pretty well topped out at about 125ft, although E-One has been building a 135-foot version. Rather than make them go higher, builders have been developing systems that are more maneuverable, with automatic stabilizers and safety systems.

Booster Tanks and Chemical Weapons

Another result of the conversion to gasoline power was that each truck or engine could carry more equipment, and it wasn't long before the pumper was combined with the hose cart and even with the ladder truck and chemical wagon.

Early chemical wagons used an alkali solution that, when combined with a small quantity of

Here's another big Seagrave from the distant past, still wearing original paint and equipment. This one is tricked out with a massive brass deck gun for throwing a heavy master stream a hundred feet or more.

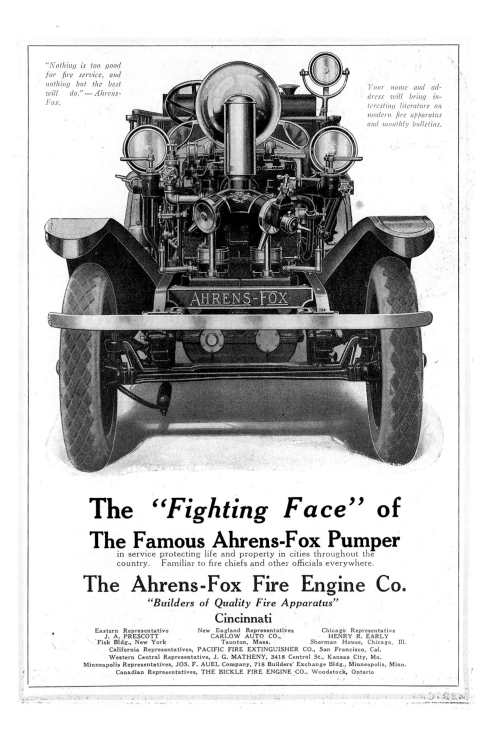

An Ahrens-Fox advertisement from 1926. "Builder of Quality Fire Apparatus."

acid, produced enough pressure to squirt the solution out of a small hand line and nozzle. These chemical wagons were good for quickly knocking down a blaze before it had a chance to take off and

were the only practical rapid method for attacking a fire until replaced by the combination pumper with its booster tank and hose reel. Then the automotive pumper came along, and "booster" tanks with several hundred gallons of water served the same function as the old chemical cars, without the fuss and bother—and with the capability of being easily recharged from any available water source. It took awhile for the conversion to be completed, and chemical wagons were manufactured right up to World War II, but for the last fifty years or so just about every engine on the street has a booster tank aboard for knocking down the little fires before they get to be big ones.

Fire engines today typically carry at least 300 gallons of water and sometimes much more; the chemical tanks with their jars of acid and tanks of alkali solution have been replaced by a simple pump, but the idea is just the same. In fact, most fire trucks today carry a booster tank and pump aboard, although they are usually both smaller than those on engines.

Pumps

As automotive power began to transform the technology of fire engine design, much thought and effort—not to mention hype—was put into competing pump designs. The three basic types—piston, rotary, and centrifugal—had been around in one form or another for a long time, but hand and steam power were far more successful with piston pumps than the alternatives. The gasoline engine, however, worked best at much higher rotational speeds than did the steamer, favoring use of centrifugal pumps.

About the same time that ladders were being radically improved, just after World War I, manufacturers battled over the relative merits of pump designs. Ahrens-Fox, American LaFrance, and Mack competed heavily for the market for engines and trucks, none of them with much modesty. The following is from a 1921 American LaFrance publication.

"**The Famous Providence Test**
"After a great deal of investigation, careful experiments and tests, it was decided that the Company should introduce the two new (piston and centrifugal) American LaFrance types at the Providence convention in August of 1916. Togeth-

The little Seagrave Suburbanite was a real hit with fire departments all around the country when it appeared in the mid-1920s. It was small, sturdy, and inexpensive enough to be just the ticket for many little towns buying their first automotive apparatus. This is one of three purchased by Morgan Hill, California, and the only one to survive intact. It has been lovingly restored by the town's fire department using parts from the other two. Below, the engineer's pump panel controls the essentials of a fire engine, the water flow from source to the nozzle. Although modern rigs seem far more complicated, the idea is still the same, as represented by this set of controls and gauges.

er with a Type 12 rotary gear pumper, the Type 12 piston and centrifugal pumping cars began the famous twelve hour Underwriter's Test, the feature of every convention of the International Association of Fire Engineers.

"It was an undisputed triumph for American LaFrance engineers. All three cars finished this racking test with a perfect score, and in their records demonstrated the marked superiority of American LaFrance motor apparatus of all designs."

Actually, while American LaFrance was congratulating itself, Ahrens-Fox and Seagrave were claiming that they were the heroes of the test. Each apparently used a carefully crafted set of criteria to support their claim, but the spectacle of these

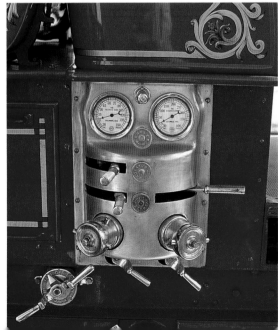

AHRENS-FOX MODEL N-S-444

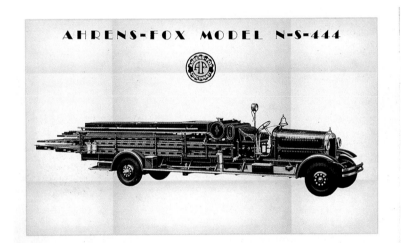

AHRENS-FOX MODEL H-44

MODEL N-S-444

Motor—5½"x7"—6 Cylinders

LADDER EQUIPMENT

50-ft. Extension Ladder.
35-ft. Extension Ladder.
35-ft. Extension Ladder.
28-ft. Wall Ladder.
24-ft. Wall Ladder.
20-ft. Wall Ladder.
18-ft. Roof Ladder, with folding hooks.
16-ft. Roof Ladder, with folding hooks.
12-ft. Roof Ladder, with folding hooks.
Total Ladder Footage, 238 ft.

LADDERS

Either trussed or solid side ladders can be supplied, but when trussed ladders are chosen it is recommended that all ladders 20 feet long and under shall have solid sides.

The usual standard complement of City Service equipment is included in the specifications of the Model N-S-444.

This model may be furnished with chemical tank or the booster pump and tank auxiliary (as illustrated above).

All models are equipped with 4-wheel brakes and booster brake auxiliary.

FIGHTING the flames—that's the business of any fire-fighting apparatus, but with Ahrens-Fox equipment it means infinitely more. Always ready for action . . . rapid mounting and dismounting of ladders . . . rigidity under stress . . . rugged construction to defy rough usage . . . modern designs . . . latest engineering features . . . best materials money can buy . . . years of specialization in the manufacture of equipment that will most successfully safeguard property and protect life . . . these are some of the reasons why Ahrens-Fox apparatus has become the standard of efficiency and preferred when only the best is considered.

Consult with us on your problems. The experience we have gained through building City Service Trucks for other cities is yours for the asking.

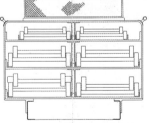

This illustration shows the modern method of "Double Bank" ladder housing and nesting. Two steel angle bars for each compartment maintain the rigidity of the ladders and decrease vibration or "whip". Ladders slide over steel guides and are held in position by locking clamps, front and rear. Always ready and within convenient reach. The banking of 238 feet of ladders is shown.

Conditions vary . . . and we are in a position to tell you how others have solved their problems.

Through the years of studying the needs of fire-fighters, designs have been modernized, more effective ways of doing things have been found, increased speed has kept pace with new demands, methods of production have changed, but our original ideal to build better equipment remains the same. We know that not only the proper functioning of this apparatus is absolutely essential under all conditions, but the Ahrens-Fox standard of mechanical excellence must be maintained if we are to continue to experience the loneliness of leadership. Study the equipment shown in this folder, then tell us your story. Particulars, specifications, quotations or proposals on request.

MODEL H-44

Motor—5"x6"—6 Cylinders

LADDER EQUIPMENT

50-ft. Extension Ladder.
35-ft. Extension Ladder.
35-ft. Extension Ladder.
28-ft. Wall Ladder.
24-ft. Wall Ladder.
20-ft. Wall Ladder.
18-ft. Roof Ladder, with folding hooks.
16-ft. Roof Ladder, with folding hooks.
12-ft. Roof Ladder, with folding hooks.
Total Ladder Footage, 238 ft.

LADDERS

Either trussed or solid side ladders can be supplied, but when trussed ladders are chosen it is recommended that all ladders 20 feet long and under shall have solid sides.

The usual standard complement of City Service equipment is included in the specifications of the Model H-44.

This model may be furnished with chemical tank or the booster pump and tank auxiliary.

All models are equipped with 4-wheel brakes and booster brake auxiliary.

AHRENS-FOX FIRE ENGINE CO.
CINCINNATI...OHIO

American LaFrance's Super Pumper, the 3,000gpm, twin-engine Metropolitan delivered to Los Angeles in 1937. The Super Pumper (officially known as the Duplex Metropolitan) used the unusual enclosed cab with accommodations for three. It used two V–12, 240hp engines driving a two-stage parallel series centrifugal pump that could put out so much water that one of these rigs could replace several conventional engines at a major fire. *Alex Matches collection.* Opposite page, top right, Moreland Brush Fire Truck, 1930. The Los Angeles County Department of Forestry bought this rig for fighting the brush fires common to California. It carries 650 gallons of water to feed a small pump and two hand lines. Moreland was one of many builders that had a good reputation, but that didn't save them from going under during the great depression. Opposite page, top left, San Jose's perfectly preserved 1937 Series 400 American LaFrance Metropolitan, hero of many fires, and one of the three most beautiful fire engine designs ever built; twelve big cylinders, a bell from a locomotive, with a steering wheel that is big enough for a ship—the classic American fire engine of all time. Opposite page, bottom, Ahrens-Fox city service hook and ladder trucks brochure. *Hall of Flame collection*

Newark, New Jersey, bought this delicious 65-foot watertower in 1936. It uses a spring-assist system to raise the tower, probably the most complex and capable ever built. Automotive power was provided by an American LaFrance Metropolitan chassis with a V–12 engine. Newark specified that the rig be painted white. *Alex Matches collection*

manufacturers all claiming superiority in the trade journals must have seemed silly at the time. Five engines participated in the twelve-hour test and on-ly one operated for the whole period without having to shut down—and that was an Ahrens-Fox.

While American LaFrance pushed its rotary gear design, all three designs had real virtues and advocates. Ahrens-Fox used a piston pumper to throw a stream to the top of New York's Wool-worth Tower in July 1917, just short of 800 feet above the rig on the street. American LaFrance's rotary was certainly a hit—they sold over 4,000 between 1910 and 1926—but their other offerings were less popular. During the same period, American LaFrance sold a grand total of only forty-one piston-pump models and a pathetic five centrifu-

gal pumpers (by 1930, the company had dropped the latter from the line).

One of the big hits of the decade was certainly the little Seagrave Suburbanite, introduced in 1923 at the annual convention of the International Association of Fire Engineers (IAFE). With a sim-ple, inexpensive 350gpm pump and a petite size, the Suburbanite was just the ticket for all the small towns around the country, and big ones as well. New York bought fifty at one time, but without the pump—they were all used as hose cars!

The Suburbanite became a kind of "low-end" introductory automotive engine for the smaller communities that were just growing into the auto-motive age. Like the other models in the Seagrave lineup, it used the time-tested chain drive, and looked just like a smaller version of the big rigs,

Five years ago, at Anchorage, Alaska, an American LaFrance Type "12," registered No. 3970, pumped continuously for thirteen days without a shutdown, except for the purpose of changing oil, maintaining a pressure of 200lbs. for this period which is equivalent to driving the car 9,400 miles at full speed practically without interruption.
—From a 1927 American LaFrance advertisement

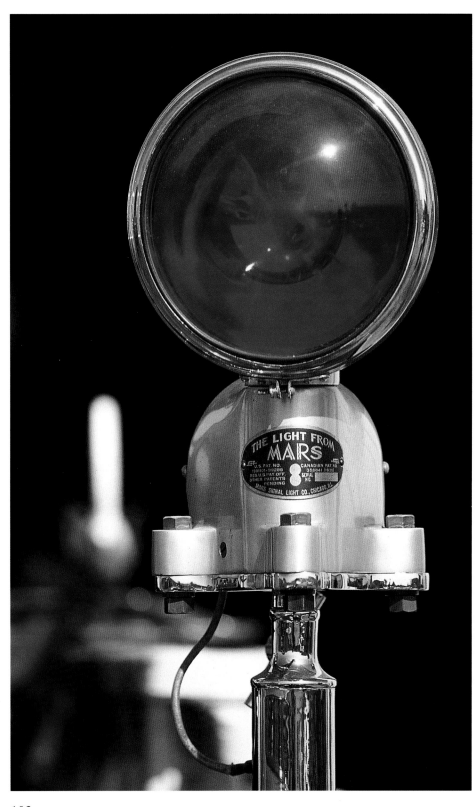

THE LIGHT FROM
MARS

too. But, unlike the big rigs, the little Suburbanite could navigate the muddy streets and roads that were a fixture of even many urban areas of the time, where the heavy pumpers were likely to get bogged down.

Like other manufacturers of the time, Seagrave built something for every size fire department. Besides the little 350gpm Suburbanite, the company built the Standard(750gpm and 1,000gpm models) and the Metropolite (1,300gpm). Some of these were still being delivered with solid rubber tires and wooden artillery-style wheels while others came with pneumatic tires and steel wheels.

Ahrens-Fox was finally delivering its immense, theatrical, adorable piston pumpers by 1915 and would continue to sell them for the next thirty-seven years, until 1952, with comparatively minor modifications. The biggest of these back then put out 1,200gpm, and achieved a perfect score at the annual IAFE convention Underwriters Test (the World Series of the fire service). That achievement helped establish the company's reputation for excellence even further, but they kept making steam-powered pumpers anyway; besides, there was still a market for them.

The early 1900s were a great time to be in sales, particularly if you happened to work for a motorized fire apparatus builder and worked on commission. The Fire Department of New York bought fifty automotive rigs in 1915 alone. New Orleans ordered twenty-two engines from the White Motor Company in 1918, paying $7,700 each; the department was converted from horses to automotive power almost instantly. But it was a bad time to be selling obsolete steam rigs or electrical drive; the latter was out of production by 1919.

Stutz Fire Engine Company was one of the many automobile builders that got into the market, but Stutz did it in a grand way. Their first design, a triple combination, was demonstrated at the annual IAFE convention at Kansas City in 1919 and was the hit of the party. The next year, the company sold a whopping thirty-five rigs to Indianapolis alone, although it took two years to deliver them all. The order included eighteen 600gpm Model Bs, seven 750 Model Cs and ten service trucks.

Coping with Depression

When the stock market crashed in October 1929, the collapse of the economy was almost universal and affected the fire service as much as anybody. Orders were canceled or were not paid for. Rigs were returned when the departments that ordered them discovered there wasn't money available to pay for them.

Most manufacturing companies tried to stay in business any way possible, and to keep employed the maximum number of workers, usually by having a maximum number of people working part time. While there were a few sales to be had, times were austere.

But despite the gloom and doom, some companies found ways to cope. And the Depression was not all that depressing all the time—there were moments of prosperity for some communities and

This old Mack was probably delivered to the Chicago Fire Department around 1920 but it was still soldiering along here on February 5th 1947 at a fire at 36th and Morgan Streets. Conditions like these are one reason that fire engines are usually engineered and built to standards higher than almost any other kind of vehicle. *Hall of Flame collection.* Left, it looks like an American LaFrance Metropolitan but it is actually a Mack from Chicago's Engine Company no. 40, with enclosed cab and midship pump, on scene at Loomis and Lake Streets on August 20, 1946. *Hall of Flame collection.* Far left, the Light from Mars was one of the few anemic warning devices installed on fire apparatus before World War II. This light is mounted on a 1927 Seagrave.

103

Despite a reputation for durability, the Ahrens-Fox pump could still occasionally blow a gasket. This one failed spectacularly on October 3, 1945, at a major Chicago fire. *Hall of Flame collection.* Right, although few water towers were built, Chicago owned a lot of them, including this one worked a "job" at the office of the Decca Record Company distributor in June of 1948. *Hall of Flame collection.* Lower right, this little 1931 REO Speed Wagon/Howe triple combination fire engine served the little community surrounding Boulder dam in Nevada and is still a member in good standing of That Dam Fire Engine Company. It was a pile of rusted junk not too long ago, before it was discovered by Steve Frady and a small group of fire fanatics who then threw money and time at it for a few years, with this result.

industries during the period. Sales for most builders dropped to about one-third or one-half of previous levels.

Ahrens-Fox, Seagrave, Pirsch, American LaFrance, and many others concentrated on building a better fire engine, and all found ways to do it. Seagrave's little Suburbanite, along with its clones from other builders, sold well to cities that needed a new engine at a bargain price.

Ahrens-Fox continued to make its big piston pumpers for a smaller upscale market, but still sold some. They continued to find ways to improve their flagship product line for fire departments that needed a rig the size of a small ocean liner.

Ahrens-Fox also started selling a smaller, less expensive rig that competed with the rest of the industry. The new line was introduced in 1935, based on a centrifugal pump of home-brewed design. The next year produced another little corvette for the Ahrens-Fox fleet, the SC (*S* for Schacht, the chassis manufacturer, and *C* for the 500gpm centrifugal pump installed amidships).

All this was not enough for Ahrens-Fox and other companies. Ahrens-Fox went bankrupt in 1938, but stayed in business for another fourteen years, although never with the vigor of its youth.

Despite—or because of—the stresses of the time, much innovation took place in the business of building fire engines and apparatus during the 1930s. American LaFrance, Seagrave, and Pirsch all managed to come up with big, beautiful, timeless designs. In 1935, American LaFrance introduced the 400 series Metropolitan, one of the prettiest fire engines ever built. Pirsch came out the same year with a 100-foot, powered aerial, and Mack started selling rigs with enclosed cabs.

Engines were bigger and more reliable. Frequently, they were twelve-cylinder; drive trains were now using shafts instead of chains. Hydraulics began to replace muscle power for some functions. The decade wasn't all that depressing.

The Diesel Era Begins

Although diesel power is accepted as a standard, basic system of propulsion today, it was only first introduced in 1939, four decades after the internal combustion engine started to displace horses in the horsepower department. Columbus, Indiana, bought the first diesel fire engine, a 1000gpm pumper built by the New Stutz Fire Engine Company. Under the hood was a straight-six, 175hp engine built by Cummins, bolted to a five-forward-speed transmission.

A detail of the spring-assist mechanism used to erect American LaFrance aerial ladders. *Alex Matches collection*

While there has been a continuous trend to modernize in some areas, some things in the fire service haven't been improved upon, despite many decades of attempts. One of these is the pike pole and another is the old "New Yorker" leather helmet that is still popular after at least 150 years. This helmet design has many virtues unequaled by the most modern designs and, although fairly expensive, tends to be a once-in-a-career purchase. Unlike the high-tech Kevlar and other modern systems, the leather helmet won't melt or crack; it offers excellent protection against heat and falling ceilings. And it looks like a fire helmet should. Lower right, here's a handsome little 1941 Ford truck converted by Howe and sold to the US Navy. Powered by a compact but powerful flat-head V–8 engine, it was a good alternative to the bigger, more expensive rigs of the time.

Figuring out how to power a fire engine is a different problem than with a conventional vehicle since the powerplant has two functions: moving the vehicle and powering the pump. Even the best gasoline engines tend to accumulate carbon deposits and have other problems when operated for long periods at a constant output, as they are when pumping water. All major cities occasionally have big fires that last for days, testing the stamina of firefighters and apparatus as well. The diesels were—and are—unbeatable for reliability.

A new fire engine normally is expected to have a twenty-year life and put on about 100,000 of the worst possible kind of driving miles. During that time, it should never need overhaul, never need to have the head removed, or any similar major mechanical work done. That is a legacy of the introduction of diesel power.

"With the old gasoline engines, we would have to do an overhaul on them by about 50,000 miles," says Don Wisinski, retired master mechanic. "But with diesel engines we never did one the whole time I was in the shop—and I started in 1970! A fire engine should get at least 150,000 miles before needing an overhaul, and that's about twenty years worth of service."

Fabulous Forties and Frivolous Fifties

During World War II, the apparatus builders stayed busy, but their output was largely going to

This 1944 Seagrave spent most of its early life in uniform, wearing Navy gray, but was rescued by Allan Gilliam and painted a good, proper, fire-engine red. Left, Marietta, Ohio, bought this American LaFrance JOX aerial in the mid-1940s. The JOX has the sad distinction of being probably the ugliest fire engine design ever. *Alex Matches collection.* Below, a 1949 Seagrave 85-foot aerial goes into action at a Cleveland, Ohio, fire in 1973. *John McCown*

government agencies instead of civil jurisdictions. Cities and towns had to make do with what they had; many dug deep into the municipal warehouses and corporation yards for anything with wheels and a pump. San Jose put a 1898 American steamer back into service and many other little cities did, too; there was not much of a choice.

But with the conclusion of hostilities, years of pent-up demand for fire apparatus were released on any available manufacturer. It wasn't just the wartime neglect of new rolling stock, but the austere times of the Depression that preceded the war that accounted for what was about fifteen years of deferred purchases. San Jose put the old 1898 American steamer back in mothballs and, as did the rest of the nation's fire departments, went shopping. After fifteen years of neglect, the sales staffs for the builders were suddenly popular again.

By the 1950s Mack and other builders were designing rigs like this handsome Type 95 triple combination—about twice the size of prewar engines with the same configuration. Firemen (there weren't any firewomen) mostly rode on the tailboard, and occasionally fell off.

Although there were plenty of rigs from the 1920s and 1930s still in service after World War II, manufacturers were building new apparatus with radical changes to the vehicles' layout. Enclosed cabs became more common, although hardly universal. As the physical size of the rigs increased, manufacturers started putting the engine amidships with the cab forward for better visibility.

Some of the innovations, however, were downright goofy. Hahn built a massive rig that had two engines and transmissions, eight wheels (two sets forward and two aft, all powered); both sets of forward wheels were steerable. The two 100hp Ford engines provided propulsion and pump power; each was started separately, then engaged to its transmission. It was a nightmare of synchronization problems, with one gas pedal and one shift lever for the dual engines, and it did not seem to work too well. Only one was built and it went to the Friendship Fire Company in Morgantown, Pennsylvania. Morgantown sent it back shortly thereafter as unsuitable, and bought something else.

New rigs appeared in towns and cities all over the continent—with plenty of variation in layout and specifications. Seagrave kept the old engine-forward style, with the siren stuck dead center where the grille belonged. American LaFrance started selling new, improved cab-forward engines. The snorkel—a boom arrangement that was adapted from the elevated work baskets used by telephone and power companies to work on phone and power lines—appeared in Chicago in 1958, then everywhere else shortly thereafter.

Water Towers Get Washed Away

Development of the aerial ladder, and later the snorkel, eliminated the need for the water tower. As ladders became longer and stronger during the 1930s, it became practical to mount large-capacity nozzles on them, allowing a firefighter to attack a fire from a well-elevated perch.

A 1958 Diamond T complete with a traditional crew member aboard, ready for a fire. Left, although its JOX ancestry is apparent, this American LaFrance Series 700 manages to be much more attractive than its wartime parent. Putting the cab in front of the engine was one of those startling innovations that look odd for a bit but offer plenty of advantages: better visibility, steering and weight distribution and enclosed seating for five. This one went to Tuckahannock, Pennsylvania, shortly after World War II.

YOUNG MEN'S
VOLUNTEER
FIRE CO.

BLAIRSVILLE, PA.

AMERICAN-LaFrance   7134-B.

The Spangler Dual, Model DH-4-200, was one of those things that are common to the fire service and that are usually called either noble experiments or harebrained ideas, depending on the observer. This monster used two engines and a very complicated synchronization and drive train that didn't work very well. The Friendship Fire Company of Morgantown, Pennsylvania, sent it back to the manufacturer shortly after delivery. *Bob Ward collection.* Right, a big Ahrens-Fox labors away in its declining years with the Chicago Fire Department. The occasion is a grain elevator fire, December 6, 1946. *Hall of Flame collection.* Far right, there was a time when enclosed cabs, like this Mack design, were considered to be quite unnecessary and money wasted on coddling firefighters. This firefighter doesn't seem to mind being coddled. *Hall of Flame collection.* Previous pages, a delivery photo of an American LaFrance Type 945 quad with a single-stage centrifugal pump and eight-cylinder engine.

This arrangement offered longer reach and better control than the water tower could deliver, and by 1938 the last had been built and delivered to the Los Angeles Fire Department. Although few were ever manufactured, most big cities had at least one, and some of these were spectacularly powerful; the Fire Department of New York bought an 8,500gpm rig from American LaFrance in 1930 that could knock a brick wall down with its master streams from the tower and two deck guns.

During World War II many commodities were earmarked as essential war materials and were unavailable for civilian use. Metals such as chromium, copper, nickel, and stainless steel were in especially short supply, so fire trucks and engines built during this era lack the sparkling chrome and polished metal of earlier rigs. This 1942 American LaFrance advertisement explains the change in finish and how the change serves the American war effort. The name "American LaFrance" was hyphenated in the early days, but now is not. Far right, top, a little Ford truck conversion. Far right, bottom, a 1953 Type L95 Mack at speed, one of three that formerly served Salinas, California.

Full Dress for VICTORY

A new color trim to replace the chrome-plate is a part of YOUR contribution to insure VICTORY. No sacrifice in performance of your fire engine has been necessary—yet YOUR cooperation has made available such critical materials as nickel, chromium, copper and stainless steel for production of armor and equipment for our armed forces.

One item—nickel—formerly used as the under plating on chrome plate trim on American LaFrance fire fighting apparatus (for 12 months production) is sufficient to properly alloy 50,000 pounds of armor plate. Yes—YOU are insuring VICTORY when ordering your fire engine in war dress.

The full productive capacity of American-LaFrance-Foamite Corporation is being devoted to defense needs and fire apparatus is being produced utilizing all of our resources and facilities, that YOU may have the protection against fire that may be needed.

The **extra** performance and reserve capacity of American LaFrance fire engines is proven in service every day. For **dependable** fire protection for YOUR community insist on American LaFrance equipment.

Snorkel Bob

Chicago pioneered the use of snorkels during the late 1950s. The snorkel is a fixture today in most departments thanks to the imagination of Robert Quinn, Chicago commissioner. Quinn had the idea of combining a commercial Pitman Manufacturing Company cherry picker with a 2-inch hose line and 1,200gpm monitor in the basket, all attached to a simple little commercial GMC 354 truck chassis.

While the first prototypes were being tested, they had several opportunities to be used on fires and on rescues and performed superbly. They quickly and efficiently helped rescue people from an elevated train platform, a school fire (in which, tragically, 92 children died), and a lumber-yard fire. The whole rig, custom built by the Chicago Fire Department shops, cost only $14,000 and revolutionized the business of aerial attack and rescue. The department bought several more and

This American LaFrance Series 800 engine served the city of Santa Clara, California. After it was retired, it was restored in a prison auto shop and is now perfect, inside and out.

found them all to be just the ticket for those nasty cockloft fires and roof rescues. Commissioner Quinn would henceforth and forever be known as "Snorkel Bob."

Even though the all-metal, hydraulic-operated ladder was introduced a decade earlier, the Fire Department of New York did not get around to buying any until well after World War II, in 1948. In fact, the FDNY was still buying antiquated

spring-loaded wooden aerials from the Four Wheel Drive Auto Company (FWD) in the middle 1950s.

Firefighting Enters the Jet Age

The trend toward cab-forward configuration continued during the sixties, along with a trend toward smaller fire companies and more diverse missions for the firefighters. For many reasons, the

size of fire companies shrunk drastically during and after World War II, resulting in fewer people doing more kinds of work. The designs of the engines and trucks were based on specifications from these smaller fire departments. A fair amount of innovation and experimentation was done with new materials and systems. As a result, the fire engine of the sixties began to look as different from its prewar ancestor as the early automotive rigs had from the horse-drawn steamers and ladder rigs.

As usual, there were a series of entertaining technological experiments that fell flat. American LaFrance, for example, came out with a pair of experimental rigs powered by aircraft jet engines in 1960, but neither got off the ground. Acceleration

was sleepy, noise was excessive, and reliability was low. It wasn't long before the gas turbines were pulled and a conventional power pack was installed.

But some of the experiments were direct hits. The snorkel introduced by the Chicago Fire Department in 1958 was a huge and immediate success. Departments from the United States and Canada acquired them as an alternative to conventional aerial ladders. Several versions evolved, including the Telesquirt, that quickly put a high-capacity appliance up above the street.

Specifications for fire engines and related apparatus began to include space provisions for equipment unknown to previous generations of firefighters, for missions they had never consid-

This handsome American LaFrance engine is forty years old but still in virtually perfect condition. The Series 800 included nine equipment storage compartments, space for 2,100 feet of 2 1/2-inch hose, a 300- or 500-gallon water tank, plus a choice of seven engines (five six-cylinder and two V–12s) developing from 175 to 305hp.

When Zoltan Szucs was eight years old he was allowed to visit the local fire station and to sit in the department's brand new 1963 American LaFrance Series 900 engine. Later Zoltan joined the department as a volunteer and learned to drive a stick shift on the same rig and rode it to many fires and emergencies. Finally the old rig was declared surplus and put up for auction. It went home with Zoltan for a high bid of $2,000.

ered. Lockers for SCBA packs and spare air bottles appeared in specification sheets, along with storage for all kinds of specialized nozzles and valves. Fire engines started appearing with forcible-entry and specialized tools that once belonged only on trucks and rescue vehicles, and lifesaving equipment that previously was the sole property of the rescue company. All this gear made for a much bigger, heavier vehicle.

The shape of things to come also included a fully enclosed cab by 1960—something of a radical and controversial development. The open cab really did have some advantages: visibility in traffic and on-scene are unquestionably better, and both are extremely important. The open cab also made

for a fast attack once the rig pulled up to the fire. The early enclosed cabs were recycled truck designs whose inventors had not anticipated the problem of wearing the old "high eagle" fire helmet inside such a space. Getting in and out of these cabs was much more awkward and took up precious time.

On the other hand, fire engines and trucks get into a fair number of traffic accidents, and the enclosed cab was (and is) an important safety device. One early version of the enclosed cab design went skidding off an ice-covered road and rolled down an embankment. Because none of the firefighters aboard were killed or even badly injured, the event made news in the trade journal

Fire Engineering. It was obvious that the crew would have been ejected and probably killed if they had been riding in an open cab.

Then there was the cold factor. Despite appearances, firefighters do not actually enjoy pain. There are few things more painful than fighting a fire on a subzero January night in Cleveland or Chicago, then chiseling the ice off the steering wheel and seats to drive back to the station. The open cab gave the firefighters a beating on the way to the fire and on the way back, too, during those cold winter responses. By the early sixties, most apparatus, even in the warmer climes, provided some overhead protection for the crews.

Diesel power began to displace gasoline in the sixties, again for several reasons. One was that

Who else but San Francisco would try putting a gas turbine engine in a fire engine? While the "jet" engine has many virtues, fuel consumption, responsiveness and acceleration are not among them. The Turbo Chief was converted to conventional power shortly after its first attempt to climb San Francisco's steep hills. *Alex Matches collection.* Far right, this big American LaFrance aerial from about 1970 is the kind of rig that is currently coming up for sale.

most commercial truck powerplants were switching to diesel, which had proved to be more economical to operate and maintain and was far more reliable and uncomplicated.

And the innovations just kept coming. Ladders 100 feet and longer became common. The wooden extension systems of prewar design went out of service, replaced by new steel and aluminum systems. The snorkel idea was adapted in several ways, including aerial platforms designed especially for the fire service, complete with large monitors and room for several people. The Stuphen Company came out with an innovative aerial tower that fit aboard a Ford truck chassis, and sold hundreds.

Pirsch, Maxim Motor Company, Seagrave, and other companies still sold conventional fire engines with the cab amidships, but the layout became less and less popular by the middle of the 1960s.

Super Pumper

Some departments experimented with extreme designs, one of which was the Fire Department of New York's Mack Super Pumper of 1965—an approximately 9,000gpm fire engine that could probably quench the fires of hell (if there is a hydrant available). This monster rig does not respond to house fires—it can knock the house over with its master stream. Instead, it was intended for the big urban conflagrations that needed large volumes of water. That master stream comes from a deck gun aboard a companion hose wagon, a similarly bloated design with a

TURN ON
BIGALARM
SAVE YOUR LIFE

AMERI

Crown fire engines like this pretty one are comparatively rare in the East. Far right, a nice demonstration of the comparative merits of aerial ladders and snorkels at a 1986 Cleveland fire. *John McCown*

10,000gpm water cannon fed by four 2 1/2-inch lines. The hose wagon carries a mile of big 4 1/2-inch supply line.

Despite the generally brisk sales (or because of them), the 1960s was also a time of turmoil for the fire apparatus industry, with companies being bought and sold—often by financiers with no knowledge or long-term interest in the fire-service industry. The ATO Corporation bought up American LaFrance; FWD sucked up Seagrave.

Additional pressure on the old guard came from upstart little companies with aggressive marketing and innovative designs. E-One, Pierce-Arrow, Van Pelt, Go-Tract, Ladder Towers, Incorporated, and other companies started muscling

123

Firefighter's turnouts have evolved in much the same way as has fire apparatus. This June 1950 advertisement for the Midwestern Manufacturing Company trumpets the safety features of their rubber Mackinaw Striped Safety Coat. Rubber coats are waterproof, but are very heavy and melt or burn if the firefighter gets a little too close to the flames. Today's firefighters wear fireproof Nomex turnouts. Previous pages, really hard-core fire engine fans always look forward to the annual fire engine migrations that normally begin about June and last until September or October throughout the West. Then, during the brush and forest fire season, huge numbers of rigs congregate on major fires, arriving in long convoys, with light bars flashing. Engines by the dozens or hundreds park, cheek by jowl, at improvised fire camps like this one on the 1992 Rogue River fire in Oregon, where anything and everything that rolls and pumps—including this elderly Seagrave—is likely to appear.

Friend of the Firefighter!

MIDWESTERN
PROTECTIVE CLOTHING

THE Original
Mackinaw Striped Safety Coat
Reduces Accidents — Saves Lives

Several years ago, Midwestern developed the *original* striped safety coat. The highly visible stripes are moulded to and vulcanized as part of the material. Thousands of Midwestern Safety Coats are serving fire departments throughout the world. Service records prove that these garments not only offer high safety visibility, but have the qualities required to withstand the rough usage required in your hazardous work of protecting life and property against fire.

Please remember that Midwestern *does not* offer a safety coat with the stripes painted or sewed on. The *original* Midwestern Safety Coat will give you trouble-free protection and service for the entire life of the garment. *Insist on the original!*

• PROTECTION
• COMFORT
• SERVICE

Largely through superior quality, design, fabric texture and advanced features, Midwestern is a leader in the field of firemen's clothing. All Midwestern specialized protective clothing have been developed from long years of experience . . . and each detail of design adds to the service, comfort, convenience and long wear of the garment.

Regardless of climate and performance demanded, you will find a fabric to exactly suit your needs in our vast selection of crude rubber and fabric materials. The softness and flexibility of the high-grade materials used is your assurance of long service and real protection. For full information on our complete line, see your Midwestern dealer. A letter to the factory will also bring you material samples for your inspection.

NONE GENUINE WITHOUT THIS TRADE MARK

MIDWESTERN MFG. CO., Mackinaw, Illinois
Manufacturers of the Famous MACKINAW Coats

About the only time you'll find a firefighter doing this is when a photographer from the American LaFrance Public Relations department is around and there is no threat of fire. The picture is intended to promote the company's brand of fish emulsion, Foamite, a product useful for fighting oil fires. *Alex Matches collection*

Chapter 5

Collectible Fire Apparatus Manufacturer Histories

Fire engines seem to generate the same kind of excitement that P-51 Mustang fighters from World War II do, and for some of the same reasons. They are both supreme expressions of the applied design and technology of their day. They both have been used in historic and heroic endeavors. Both are vehicles and move with speed and style. But a P-51 Mustang costs a couple of million dollars—and you can put a nice big red fire engine in your garage for about $2,000 . . . maybe less.

The last few years has seen a huge interest in private collection of fire apparatus. A surprising number of persons have one, or sometimes many, rigs. They can be surprisingly easy to find and cheap to buy. The hard part is usually finding a parking space. Ed Hass, president of the Ahrens-Fox Fire Buffs' Association, for example, bought a 1953 Ahrens-Fox from New Milford, New Jersey, for only $1,500. Of course he has had to put a little more into it over the years—new paint, a bit of engine work, and then there is the time he had to have it trailered back from Wyoming. How much does it actually cost to *own* a fire truck? Well, as Hass says, "How much have you got?"

Preservation of antique apparatus happens in many ways. People buy rigs at municipal auctions, normally for little money. These rigs are the newer, borderline collectibles. Then, there are the many professional and volunteer departments that keep and use old apparatus. There are many private collectors, some rich, most not, who own one or more rigs. Many museums are dedicated to the fire service: most large cities seem to have one, and there are specialized collections as well. Then there are the muster teams—groups of men and women, in and out of the fire service, who enjoy maintaining both the equipment and the techniques of earlier times on the fire ground. Many own and operate elderly rigs, usually purchased for a relative pittance.

For example, San Jose's local muster team bought a nice 1956 American LaFrance 100-foot aerial ladder truck for $2,500. The truck was a reserve rig for a neighboring city that maintained it mechanically if not cosmetically. The ladder had been recently inspected and certified. The rig got a new engine about 20 years ago, a big diesel that is still strong and ought to last another fifty years. About the worst thing wrong with the rig is that the paint is oxidized, but at least it is red and not that vile lime-yellow. It has a stick shift rather than an automatic transmission, but at least the power steering works. The rig was cleaned and polished and added to the fleet of rigs the muster team owns.

A collector might sneer at this rig because it has been re-engined, but for somebody who wants to impress the neighbors or who wants to prune trees in style, this kind of vehicle is perfect. And while this ladder truck is not suitable for taking all the neighborhood kids out for ice cream, there are many suitable vehicles available for about the same price all over the country. Of course, they guzzle

1937 Pirsch All Power aerial ladder truck. Pirsch sold this pretty 85-foot aerial to its corporate hometown, Kenosha, Wisconsin. Pirsch was the first builder to power the mechanisms that raise, extend, and rotate the aerial ladder, with mechanical and hydraulic assists. Like many aerials, this one includes a ladder pipe mounted on the ladder tip, a kind of rigid fixture for a nozzle that doesn't need muscle power to control. This rig still wears its original paint and lettering, and still carries its original ground ladders and rescue gear. This vehicle is part of the Hall of Flame collection.

gas and parts can be hard to find, but you weren't going to commute in the thing anyway—or were you?

Several organizations work to preserve old fire engines and trucks. One is the Hall of Flame in Phoenix, Arizona; it is an immense collection of over 100 rigs. Another is the many muster teams around the country, which restore old apparatus and also compete in historic fire activities like bucket brigades and hand pumping competitions. Then there's the Society for the Preservation & Appreciation of Antique Motor Fire Apparatus in America (known affectionately as SPAAMFAA to its initiates). SPAAMFAA is a large organization with its own research library; hundreds of its members own their own rigs. Finally, there are many fire departments around the country that just never dispose of particular rigs. The department in Klamath Falls, Oregon, kept their faithful 1930 American LaFrance pumper. While most are driven only occasionally in parades, a few antique trucks and engines still chase fires.

Collecting vehicles of all kinds has become popular over the last few decades and worn-out, derelict cars and trucks that would have been candidates for the scrap yard are now resurrected to

Hall of Flame

Many excellent museums dedicated to fire engines and firefighting exist in North America and Europe, but the largest is in Phoenix, Arizona, where the personal collection of the late George Getz, Jr., is now on display. Getz once made the mistake of mentioning to his bride that he rather admired old fire engines. Mrs. Getz decided to carve off a little of the grocery money to acquire one for his birthday. The deed was done and primed the pump of desire in Getz, who spent millions on the quest, seeking out and buying up individual vehicles and the entire collections of others. The resulting collection is now the property of a foundation set up to maintain and exhibit fire-related memorabilia.

The Hall of Flame is the largest museum of firefighting apparatus in the world. The compound contains several large industrial buildings stuffed with almost every variety of rig built from 1725 to 1960. There are about a hundred of them—pumpers, trucks, water towers, aerials—most restored but some original. There are dozens of helmets, stacks of photographs and resource materials, and all sorts of related memorabilia.

As the museum's director, Dr. Peter Malloy, explains, "We have a wider collection base than other fire museums. We collect from all over the United States and England. We have the size and resources to be the major depository of objects, photography, printed materials, and manufactured items like apparatus. We may get into archiving things like company records and journals. Our goal is to collect, preserve, and display objects relating to the history of firefighting in the English-speaking world."

Each year, about 25,000 people visit the museum. Many visitors are firefighters, some making a pilgrimage from far-off lands, others are school children from local Phoenix schools. Besides the historical interest of the old rigs, the Hall of Flame has an education program and full-time staff education curator to provide fire safety programs. There is a gift shop, too, with just about anything a fire buff could desire, including a frame for a car's license plate that reads "My Other Car is a Fire Truck."

The 100 rigs in the collection include hand pumpers from all over the country, and some from Europe and Asia. There are motorized rigs from about 1908 to 1965. Many of the motorized rigs were in poor shape on arrival and had been candidates for the junkyard. These lost causes get the attention of Don Hale, the Hall of Flame resident craftsman and curmudgeon. Hale spends many months disassembling and rebuilding rigs in his large shop, sometimes imagining what a missing component might have looked like. Then there are the problems like his current project, a ladder rig whose geared turntable has had all the gears ground off and whose front wheels are missing.

new glory. In fact, some people invest in collector cars the way others buy and sell stocks and bonds or pork bellies. Within this grand movement is the fire apparatus collector cult, a community of people with large yards, tolerant spouses, and an affection for red paint.

Rather like the horse set, each breed has its own aficionados. Motivation for each varies widely, but the principal motivations seem to be availability, local service, and personal taste. For example, some folks think the Ahrens-Fox is just the bee's knees while others consider it a board-certified ugly duckling. Ditto, of course, for the American LaFrance Metropolitan, Seagrave Suburbanite, or Mack 95. Some of the most popular brands in one part of the country are unknown in others—Buffalo, Sanford Fire Apparatus Company, and Prospect stayed pretty much close to home. But Seagrave, Ahrens-Fox, American LaFrance, Mack, Pirsch, and a few others became fixtures in departments around the country, then in private ownership after retirement.

This chapter provides a thumbnail sketch of each manufacturer, just in case a rig shows up for sale for an irresistible price and you elect to join the club of owner-operators.

Ahrens-Fox Fire Engine Company

Of all the juicy, delectable collectibles, the glorious Ahrens-Fox front-end piston pumpers of the 1920s and 1930s certainly take the cake. These engines are so theatrical, so beautifully designed and built that a surprising number are still owned by the same fire departments that bought them seventy years ago, took them out of front line service half a century ago, and have stored them ever since, trotting them out once a year or so for a parade. Amazingly, at least eight are still in service.

The Ahrens-Fox Fire Engine Company originally made steam apparatus but gradually converted to gasoline power as the old technology gave way to the new. The original Ahrens Manufacturing Company had been absorbed in the great merger that formed American Fire Engine Company in 1891, but by 1904 Chris Ahrens was ready to get back into business again, this time with his sons John and Fred plus his sons-in law George Krapp and Charles Fox. They called the new business the Ahrens Fire Engine Company and sold a steamer called, traditionally, the Continental. Charles Fox took over as president in 1908 and changed the name to Ahrens-Fox. They built their last steamer in 1912. The big front-mounted piston pumper, the foundation of the Ahrens-Fox legend, was introduced in 1914 and variations on the theme were built right up until 1952.

SPAAMFAA maintains a roster of rigs that members own and a surprisingly complete sequence of registration numbers for this one man-

The few, the proud, the Ahrens-Fox piston pumper. This 1947 delivery photograph for Carlisle, Pennsylvania's, shiny new rig records one of the last of the old breed. This is one of the most valuable of collector fire engines and if one should show up at a municipal auction you are quite unlikely to snatch it up for a grand or two. *Hall of Flame*

In the dusty recesses of many a fire house are scenes like this one: a grand old Ahrens-Fox, its fire fighting days long over but too good for the scrap heap.

ufacturer survive. According to a report in *Enjine! Enjine!* in 1992, there are 264 Ahrens-Fox rigs held by SPAAMFAA members alone; that is around 22 percent of everything the company built, a far higher percentage than that for American LaFrance (670 rigs owned by SPAAMFA members), Seagrave (420), Buffalo Fire Appliance (310), Pirsch (78), Maxim (65), Sanford (62), and Howe (56). Unfor-

tunately, data for the other manufacturers isn't available to provide a comparison percentage of surviving rigs but it is quite likely that Ahrens-Fox is the best preserved of all motorized fire apparatus.

Actually, Ed Hass (founder in 1970 and still chief mechanic of the Ahrens-Fox Fire Buffs' Association) estimates that about one-third of all the rigs that Ahrens-Fox ever built (600 to 700 of the

1,600 vehicles manufactured between 1911 and 1952) are still around somewhere. Besides the SPAAMFAA members, the things keep turning up on farms, where they are used as utility vehicles, or sitting in people's yards, sometimes under canvas or dirt and rust.

Hass got hooked on Ahrens-Fox as a kid when his father took him to visit the Fairview Fire Department in New York state where an Ahrens-Fox "quint" was in honored residence. This was a one-of-a-kind rig (well, most rigs are at least a bit unique) that combined a 1,250gpm piston pump with a 70-foot, electrically-raised aerial ladder (and for 1939 that was the cutting edge of technology), a 1,197ci displacement engine, and a 43-foot chassis with duals at the rear.

Ahrens-Fox built their first piston pumper in 1911 and the last in 1952, a total of about 900 piston-pumper rigs, far fewer than the more popular Seagraves and American LaFrances. These huge, complicated, expensive engines were the pinnacle of the classic breed of fire engine. Not every city had an Ahrens-Fox; it was the Duesenburg of the apparatus world during the 1920s, 1930s, and 1940s.

Ahrens-Fox was an innovator, the first to put the hose on the same vehicle with the pump (1892), first with the booster tank in 1916, double-stacked ladders in 1923, and tied with the Stutz Fire Engine Company in 1919 with the first application of shaft drive to the rear wheels.

The company built the 500 series, a smaller, more conventional and affordable pumper, during

Ahrens-Fox is considered one of the most valuable collector makes, and these big, classic front end piston pumpers can sometimes go for over a hundred thousand dollars. This vehicle is a 1930 Quad from the Hall of Flame collection.

Oil quantity on an Ahrens-Fox V–12 is indicated by this little float gauge. Looks like this one needs to be topped off—twenty quarts or so should do it. Lower right, the drivers' position on an Ahrens-Fox is high off the ground, almost like sitting on a stage coach, but with a lot more horses to manage. The instrumentation for fire engines remained primitive for many years, until after the Second World War. A driver was expected to know how the rig was doing as much by ear and by the feel of the vibrations through the wheel as anything else.

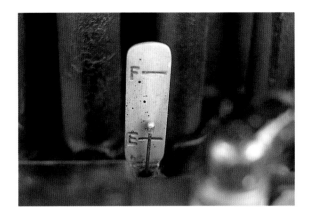

the Depression. But during the life of the company, the company built only about 1,600 rigs of all types.

Pat Madigan joined the Seattle, Washington, department in 1942 and his first assignment was on a 1926 Ahrens-Fox. "I was a rookie on that rig and went to my first fire on the Ahrens-Fox," Madigan says. "When it was pumping, it would bounce up and down like a son-of-a-gun—but it was a good pumper. It was a son-of-a-gun to drive, too! It had a small steering wheel and that big engine and pumper up on the front made it nose heavy . . . you really had to work at that little wheel to make it turn a corner. But it would go like a bat out of hell! Here in Seattle we could make all the traffic lights go red so we could drive down the main street at 65 miles per hour. I've done it!"

Bob Ward, current president of SPAAMFAA, is an Ahrens-Fox fan and has been since he was a kid in the 1920s. Of course, the large Ahrens-Foxes were and still are the most gaudy fire engines on the street, with their big exposed piston pumps and huge glittering air dome. Ward didn't get to actually own an Ahrens-Fox until later in life, but he liked the first so much that he bought a second to have a matched set. There were some problems, though; in order to get them in the garage he had

to take the roof off and add three courses of blocks to add the necessary clearance. As he explains:

"When I was growing up, my hometown, Norristown, Pennsylvania, had three 1923 Ahrens-Fox engines that were paraded every year. Then, in 1966, a friend who knew I liked them called to say that there was one for sale in Canton, Ohio. I bought it sight-unseen—for $1,125—and had it trailered here. Everything was still on the rig—the stretchers, the nozzles, everything. The neighbors all said, 'Boy, you're nuts! When can I have a ride?' We had a lot of fun with it—so we bought a second Ahrens-Fox, an 85-foot aerial from Worcester, Massachusetts. I bought that one sight-unseen, too.

"I learned about then that the cooling system holds fifty-six quarts of coolant and the engine takes sixteen quarts of oil—plus fourteen more quarts of oil for the pump! When you change the oil on one of those things you just about need a *barrel* of oil!

"They used to be expensive and they are expensive now. It used to be said that an Ahrens-Fox cost a dollar a pound, and that's just about right. In the 1920s and 1930s, a piston pumper weighed about 13,700 pounds and this is what they sold for.

"Everything about an Ahrens-Fox is big; the engine weighs a ton and so does the pump. The pistons are 5 1/2 inches by 7 inches stroke. The cylinders were each separate and exposed; you can watch the valves operate up and down. The vehicle is twenty-five feet long and you sit back thirteen feet from the front!

"It will pump a thousand gallons a minute, all day. It was nothing for an Ahrens-Fox to last for thirty or forty years in service with a city."

Ward has a lot of company in the Ahrens-Fox fan club, and you won't find many of the big piston pumpers going for $1,100 anymore. But there are still plenty around, and many of the more conventional engines as well. Ed Hass bought his from New Milford, New Jersey, for $1,500. But most of these are now too pricey for anybody but the rich collector or investor.

Although prices fluctuate in any market, approximate values at this writing for an unrestored Ahrens-Fox piston pumper are around $40,000 with the renovated rigs going for up to about $140,000 (tax, dealer prep, license extra). In spite of their collector value, some still earn their keep on the street. One, according to Ed Hass, was used by the volunteers of Oceanside, New York, until about 1988. But, amazingly, the town right up the road, Baldwin, New York, still owns and operates their faithful 1951 piston pumper in first-line service. The towns of Jefferson and Turtle Creek, Pennsylvania, both have Ahrens-Fox piston pumpers in first-line service. Ed Hass believes as many as eight may still be on line, although some of these have not been called on to go to a major burner in many years. Certainly, many departments have held on to them, stashing them in warehouses or just in the back of a firehouse to collect dust. Peter Malloy, director of the Hall of Flame explains their popularity.

"They were more individually crafted than other makes and they were a quality piece of apparatus. Even today, they may have endured years of abuse and neglect, slowly moldering away, but you can still see the quality of their engineering and fabrication. Ahrens-Fox made almost all their own components—they had incredible horizontal integration.

"The Ahrens-Fox piston pump was really an obsolescent design, based on Chris Ahrens' expertise in piston-pump design. It is a beautiful pump, very nicely made—but that's one of the major reasons the engine was so expensive. It cost a fortune to build the thing. Its major advantage was its ability to draft water from a pond or cistern. You could pump for either pressure or volume, and that provided a lot of flexibility. But the centrifugal was the pump that was best matched to the gasoline engine, and I always thought Seagrave had the best design.

"I always thought Ahrens-Fox was over-rated by the fire buffs because they're so beautiful. I guess that's true . . . if you've got a mechanic and a parts man following along behind you."

Ahrens-Fox may be famous for those big, gaudy, piston pumpers, but they made a more diversified line of engines. A smaller quad—a four-feature engine with pump, hose, booster tank, and ground ladder—with a rotary pump mounted amidships came along in 1929, christened the Skirmisher. Another model hit the street in 1930, the Model V, with a 500gpm pump.

It used to be said that an Ahrens-Fox cost a dollar a pound, and that's just about right. In the twenties and thirties, a piston pumper weighed about 13,700 pounds and this is what they sold for.
—Bob Ward, president of SPAAMFAA, owner of a 1923 Ahrens-Fox

These rigs were part of a marketing plan to appeal to the lower end of the market (the Depression made that the only end of the market) with a more economical, technologically less sophisticated product that had a chance of being sold to somebody. Three other lines were introduced during the 1930s, including the SC with an Ahrens-Fox designed centrifugal pump and the glorious HT piston pumpers that slid down the ways in 1937.

The heyday of Ahrens-Fox was during the 1920s and 1930s, when its big piston pumpers were (as Ed Hass says) the Rolls-Royce of fire engines. The Depression, however, took a lot of steam out of the company, and by 1938 the firm went broke. It reorganized and stayed in business,

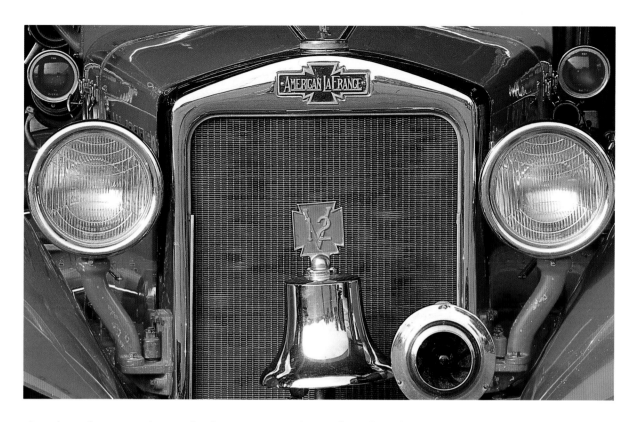

The 1937 American LaFrance Series 400 Metropolitan. This is preservation rather than restoration. Although the paint on the fenders has been thinned by the decades of washing and polishing, this rig should be left pretty much as is—although it would be nice to have use of second gear and the leaky fuel line is a bit scary.

though, and continued to make fire engines until 1952 when it finally went under and out of business. For a while, the company reverted to the same practice of many other builders employed to keep costs down, they used conventional truck chassis as a foundation for fire apparatus assembled of mostly off-the-shelf components. Ahrens-Fox even offered a midship centrifugal pumper, just like everybody else—and it looked just like all the others, too. Many Ahrens-Fox fanatics don't consider the products of this era quite kosher—but they don't complain much to Ed Hass, the Ahrens-Fox guru because he owns one of these rigs.

One interesting part of the Ahrens-Fox story is that when the company went out of business, Curt Nepper, a former employee, purchased the Ahrens-Fox name, the company records, and the entire inventory of spare parts and patterns. Nepper saw a need and intended to fill it. He has been the sole source of parts since 1952 and it is still possible to buy a somewhat new, although dusty, gasket or bracket for that 1937 piston pumper out in your garage.

American LaFrance Company

If you are looking for an entry-level fire engine, the most likely candidate in most parts of the country is a twenty- or thirty-year-old Alfie. For all its long existence, American LaFrance was a prolific builder of apparatus in all shapes and sizes, with an aggressive pricing policy that helped them dominate the market. Few firehouses around the country have not been home to at least a few American LaFrance engines or ladder rigs; some have known almost no other make. That means that there are more of them coming out of service and up for auction or more that are already in the used apparatus market.

American LaFrance was born from one of the many conglomerations of smaller fire manufacturing companies common around the turn of the century. The firm had its origins with several independent firms, one of which was the LaFrance Manufacturing Company. Truckson LaFrance founded the company in 1872 to compete in the market for steam pumpers. LaFrance had worked for the Elmira (New York) Union Iron Works, a builder of locomotives and stationary steam appa-

AMERICAN-LaFRANCE

ANNOUNCES
UP-TO-MINUTE APPARATUS

The most important achievement in motor fire apparatus building ever announced to the Fire Chiefs of America.

AMERICAN-LaFRANCE AND FOAMITE PROTECTION
A Complete Engineering Service
For Extinguishing Fires

The New Apparatus will be on exhibit at the 57th Annual Convention, Booths 1, 2, 3 and 36-47

A 1929 American LaFrance advertisement.

ratus, but in 1871 he was granted several patents on improved pump technologies and went into business on his own. While other steamers used piston pumps, LaFrance invented an alternate rotary pump and a rotary steam engine that were important improvements to the technology of the fire engine.

In 1880, the firm began building apparatus strictly to order and changed its name to the LaFrance Fire Engine Company. In 1884, LaFrance offered a piston pumper and developed a full line of this system to complement the company's proprietary rotary-gear pump. By 1900, the company had built about 500 steamers.

In 1891, the Ahrens Manufacturing Company, plus the firms of Silsby, Button, and Clapp & Jones formed the American Fire Engine Company. A further consolidation occurred ten years later, in 1901, when American merged with LaFrance and seven other small, obscure, builders of steam fire engines to become the International Fire Engine Company. Three years later this company changed its name to the American LaFrance Fire Engine Company.

During this time, it was obvious to most manufacturers that the era of automotive power was upon the industry, and they experimented with and adopted both electrical- and gasoline-powered apparatus. American LaFrance's first automotive rig was delivered in 1907, a 30hp chemical wagon that went to Boston, but it was not accepted and was returned to the company. Three years later, in 1910, they tried again with a combination chemical and hose wagon.

While these first efforts used gasoline power for propulsion, they certainly didn't use the engine to drive the pump; that would come in 1911 with Type 10 and Type 12 engines offering 500gpm and 750gpm output from a small rotary-gear system. American LaFrance made its final hand pumper in 1910 and the last steamer went out the door in 1914. The centrifugal pump came along in 1916, followed by piston pumps in 1914. These early efforts may have been small, but they were tidy little vehicles and were clearly the shape of better things to come.

The first really successful American LaFrance automotive system was the Type 31 tractor, a 75hp design that could tear down the highways at a blis-

tering 25mph. American LaFrance was still essentially a custom builder, producing to specification, but the market asked for a wide variety of apparatus. And one of American LaFrance's most successful during this time was its innovative spring-loaded ladder, hauled to the fire ground behind a Type 31 tractor.

In 1915, the little Type 40 engine came along with a choice of 250 or 350gpm delivery. By 1922, another, bigger rig came out the door, the Type 75 with a 150hp engine and a 750gpm pump. The 75 was usually delivered as a triple combination: booster pump, booster tank, and volume pump.

The first Metropolitan came along in 1926, a big, substantial vehicle that still used chain drive. These vehicles were part of what American LaFrance called the Series 100. The Master—also known as the Series 200—was introduced in 1929, the 300 in 1933, the 400s in 1935, and the 500s in 1938. A Series 500 chassis carried the

mammoth 125-foot American LaFrance aerial ladder rig purchased by Boston in 1941.

While there may not be many unattractive fire trucks and engines around, the most drop-dead ugly design ever delivered to the fire service is probably the American LaFrance JO/JOX aerials manufactured during World War II. This particular design has the lines of a squared-off whale and all the glitter of a brick. Most were plain, devoid of the usual chrome and shiny paint; pinstriping and gold leaf were for other rigs, certainly not the unadorned JOX. Happily, most of these have been remanufactured as washing machines and razor blades, with not many tears shed for their demise.

Series 400 Engine

If American LaFrance's JOX was among the more repulsive vehicles to ever disgrace the fire ground, this same company was responsible for one of the most beautiful, the Series 400 Metro-

S an Jose's 1937 American LaFrance Metropolitan. Ownership of such a rig can produce all kinds of dilemmas when, for example, choices have to be made about repainting a weathered and cracked finish that also contains gold leaf and pinstriping superior to anything that could be applied today.

politan. Although there may be some competition for the most beloved engine to ever fight a fire, the Series 400 is certainly in the top ten, maybe in the top two. When people think of a classic fire engine, they will be thinking of either an Ahrens-Fox piston pumper or the Metropolitan. The Series 400 had two major variants; the most attractive of which is the Senior Metropolitan, of which about 150 were manufactured. While some came off the line as ladder trucks, hose wagons, aerials, quad combinations, and a matched set of special-purpose rigs built for the Los Angeles Fire Department, most were triple-combination Metropolitan pumpers.

Although just about every fire engine is somewhat unique and custom built, the Series 400 had a set of characteristics that appealed to firefighters when the rigs were new and are just as appealing to fire buffs today. It was a big, heavy, loud, powerful, fast engine; most were powered by a huge V–12, often devoid of a muffler. The unique and identifying feature of the 400 is the placement of the pump directly astern of the powerplant, just forward of the cab. Although some of the smaller 400s had straight eight-cylinder powerplants manufactured by Lycoming and pumps placed conventionally behind the cab, most placed the pump in the cowl. The pumps, like just about everything else on the chassis, were delivered to the specifications of the buyer. Some preferred rotary-gear pumps, others the centrifugals. The type and size of the pump determined the identification of the rig.

The Series 400 was introduced in 1935 as a medium-sized engine with a straight eight-cylinder

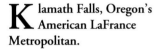

K lamath Falls, Oregon's American LaFrance Metropolitan.

Medicine Hat, Alberta, was just one of many Canadian cities that bought American LaFrance rigs. This handsome pump and hose car from the 1930s, seen new here in its delivery photograph, would be a valuable collector's vehicle if it still exists today. *Alex Matches collection.* Lower left, American LaFrance called this beautiful 1936 variation on the Series 400 Metropolitan theme the Protector. Its V–12 engine only delivered 150hp, but that was enough for the 600gpm single-stage centrifugal pump. The enclosed cab and dual rear wheels were rarities at the time, as was the glittering white paint job. *Alex Matches collection*

Lycoming powerplant and a choice of pumping systems, rotary-gear or centrifugal. Although known as a Series 400 design, this junior model was essentially quite different from the senior model that followed. These smaller rigs placed the pump conventionally behind the cab. Depending on the pump capacity, these were called the Scout (500gpm), Protector (600gpm), or Invader

If you can find an American LaFrance engine in restorable condition it is likely to be quite costly. Then comes the problem of rebuilding wooden wheels.

(750gpm). This smaller 400 was produced from 1934 to 1938, at first with just the straight eight-cylinder, later with the smaller V–12 J engine manufactured by the Auburn Automobile Company.

The Senior Metropolitan may have been built around a truck chassis but its styling suggested a much racier parentage. It has the huge, elegant hood and vast fenders of the ostentatious Duesenbergs and Cords, the big touring cars of the time. This was no mere utility vehicle but a bold, stylish machine intended to be noticed and admired. Lawrence, Massachusetts, bought the first one, a 1,250gpm fire engine built in 1934 and assigned the factory register number 7700.

Under the hood, the big E model V–12 (introduced in 1931) looks like it could propel a freight train locomotive or maybe an aircraft carrier—although it only generates 240hp. The powerplant needed seven *gallons* of 50-weight oil to fill the crankcase. Although the engines varied a bit, a typical displacement was 754ci; cylinder bore was nominally 4 inches with a piston stroke of 5 inches. Some of the later engines used a powerplant manufactured by Aburn and had an L series—a designation that endured from 1936 until the 1950s.

The first Metropolitans used the same chain drive system to the wheels that had been a part of

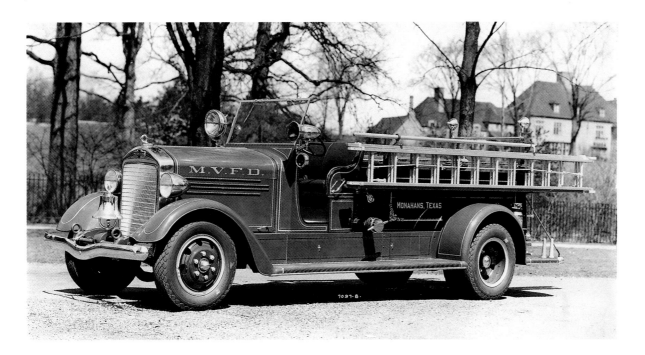

An American LaFrance Scout about to be loaded aboard a rail car and shipped off to its new home, the little Texas community of Monahans. The Scout is a 500gpm engine with a rotary-gear pump powered by an eight-cylinder engine developing 125hp. *Alex Matches collection.* Below, this was called the Type 1500 by American LaFrance, no doubt because its rotary-gear pump allegedly delivered 1,500gpm—a *lot* of water. The V–12 in this one was rated at 240hp. The rig went off to Yonkers, New York, in the mid-1930s.

the technology for the previous thirty years, but American LaFrance soon converted to the now-conventional driveshaft and differential power-train. The standard Metropolitan chassis had a 180-inch wheelbase, rolling on huge wheels and 24-inch tires in size 38x7. Shaker Heights, Ohio, bought a special one, though, with a tremendously long wheelbase that they used as a quadruple combination: booster pump, volume pump, tank, and ladder. Other unusual variants include American LaFrance's first hydraulic ladder, a three-section aerial that went to Union City, New Jersey, in 1938. Another American LaFrance first, the 100-foot aerial, went to Annapolis, Maryland, the same year.

Los Angeles bought four rather odd Metropolitans about this time—each with two V–12 engines and two 1,500gpm pumps for a 3,000gpm rig. These engines were intended to each replace three conventional 1,000gpm rigs at large fires.

Three pump capacities, 1,000, 1,250 and 1,500gpm, were offered in the fire department's choice of centrifugal or rotary-gear design. Each of these, and the pump design as well, was identified by the type specifications on the data plate. American LaFrance identified a 1,000gpm pumper with a rotary-gear pump as a 400-RB; a 1,250gpm was

a 412; and if it had a centrifugal pump, it was designated 400-CB.

In 1937, San Jose bought one of these beautiful Metropolitans, a 1,000gpm model wearing registration number 7758. On arrival, it was designated Engine 1 and occupied the place of honor in Station 1 downtown. Firefighter (and later Fire Engineer) Loren Gray joined the department in 1942, was awarded engineer-badge number 1, and drove the American LaFrance 400 for about five

The neighbors all said, "Boy, you're nuts! When can I have a ride?"

—Bob Ward

143

"Our new aerial ladder truck handles like a pleasure car," says Chief W. D. Thomas.

BINGHAMTON, N. Y. receives first of the new AMERICAN-LaFRANCE Aerials

The new American-LaFrance cab-ahead-of-engine aerial has exceeded every expectation for performance. Replacing a city service ladder truck, the 85-foot aerial ladder is new to the men, but they handle it like veterans with years of aerial experience.

Short overall length—easy handling in traffic—with ability to make short turns in narrow streets, has proven the superiority of these new aerials.

The aerial ladder built of high-tensile corrosion-resisting steel, with full hydraulic power, provides fast, safe and sure operation. For rescue or water tower use, the American-LaFrance aerial has no equal in the fire service.

RECENT AERIAL DELIVERIES HAVE BEEN MADE TO

Lawrence, Kansas
Liberty, New York
Salisbury, Md.
Kansas City, Mo.
Fort Worth, Texas
Kansas City, Kan.
Marion, Indiana
Duquesne, Pa.

AMERICAN-LaFRANCE-FOAMITE CORPORATION
ELMIRA, NEW YORK, U. S. A.
LaFRANCE FIRE ENGINE-FOAMITE LIMITED
TORONTO, 9, ONTARIO, CANADA

This American LaFrance advertisement from 1947 shows the firm's 85-foot aerial ladder fully extended.

years. "It had a 4-inch straight exhaust pipe," Gray recalls, "none of this muffler stuff! Well, we had an underpass on West Julian Street, and whenever we'd go through there I'd always gun the engine—the blast of noise in that little tunnel would just about blow the guy riding on the tailboard right off the rig! It had plenty of power and people said

you could hear it coming from a mile away. You get up behind that wheel and the hood looked like it was about twelve feet long. It was a fun rig to drive and I'm proud to have been on it."

The Metropolitan, like every other rig of the time, had an early form of power steering—muscle power. The wheel is big and it takes brute force to maneuver at slow speeds. The transmission is five speeds forward, no synchromesh. Drivers had to learn to double-clutch before they could drive without grinding gears. It wasn't built for weaklings. When Ralph Bernardo, who was a small Italian guy about 5 feet, 5 inches, came into Station 1's department and was assigned as a relief driver on Engine 1, he took a lot of kidding. There were suggestions that maybe blocks needed to be attached to the pedals for Bernardo's stubby legs. A wooden box was placed alongside the running board by the driver's side to help Bernardo climb aboard. There were comments about his ability to see over the wheel without a booster seat. Replies Bernardo,"I liked to drive it—and the job had compensations. You got to stay warm and dry while you watched the pressure on the pump panel while the guys on the hose were eating all that smoke. I decided that I wanted to drive—and that's what I did for twenty-nine years."

When Station 1's Don Wisinski, a now-retired battalion chief and master mechanic, once tried to hot-rod with a Metropolitan, he laid a patch of rubber in front of the station that took him four hours to remove. That was when the rig had already been retired to the reserve list. Even then, the department used the Metropolitan for its annual hose tests, and it pumped away for days on end without complaint.

San Jose's big 400 served its twenty years and retired to its golden years as a reserve rig. Today, it is a star performer for the muster team. It has never had its engine overhauled in its fifty-five years. The 400 still runs strongly and could go to a fire and deliver water with the best of them. If anything, it is a bit overpowered. When contemporary firefighters try to drive it, many have trouble with the clutch. It has never been restored—no restoration was needed, although the paint has been polished away in some spots by years of washing and chamois wipe downs. The elegant gold leaf and pinstriping is all as it came from the factory,

[The 1926 Ahrens-Fox] would go like a bat out of hell! Here in Seattle we could make all the traffic lights go red so we could drive down the main street at 65 miles per hour! I've done it!

—Pat Madigan, former Seattle firefighter

SPAAMFAA

The Society for the Preservation & Appreciation of Antique Motor Fire Apparatus in America (SPAAMFAA) was formed in 1958 and now has about 2,700 members who own about 2,500 rigs. "SPAAMFAA is a wonderful organization where people of all classes can come together on a common ground," says Bob Ward, association president. "Doctors, lawyers, blue-collar workers all own fire engines and belong to SPAAMFAA."

The organization sponsors two annual meetings: one in Syracuse, New York, and the other in a location hosted by one of the forty-eight chapters. These meets are the major musters of the year for the fire engine community and the best place to see living, breathing, pumping engines moving water from one place to another. But at any of the pump-ins in the East and Midwest you can enjoy the sight of sometimes dozens of rigs congregated along a pond, river, or lake, drafting water and throwing master streams high into the air.

SPAAMFAA's newsletter is worth the $20 annual membership fee. *Enjine! Enjine!* is a nicely done, highly informative, and entertaining quarterly with authoritative articles on specialized bits of apparatus history, muster activities, and ads for rigs and accessories. An advice column answers questions from the mechanically perplexed: Should the voltage regulator on my American LaFrance 700 series be grounded to the chassis? Do you use additives in your gasoline? How should you care for an oil filter made up of metal screening?

But the best part of *Enjine! Enjine!* is the Fire Flea Market section, which includes ads that make you want to make some room in the garage and have one of those meaningful chats with your spouse. For example, these two ads from a recent issue sound appealing:

FOR SALE

1939 Ahrens-Fox VC-9071 500 gallons-per-minute midship centrifugal pumper, ex-Munster IN. Completely original, has full equipment, just 4,900 miles. Excellent paint and chrome. Must sell. Price $6,000 or best offer. . . .

1946 Seagrave Open Cab Pumper and 1946 American LaFrance crew cab pumper. Both V-12s, good original condition with excellent tires. Asking $3,000 each. . . .

For more information or to join the organization, write to SPAAMFAA, P. O. Box 2005, Syracuse, NY, 13220.

and while there may be someone who could duplicate it today, the muster team likes it just the way it is.

Dana Martin, publisher of *The Visiting Fireman*, owns a 1926 American LaFrance that once served Grand Rapids, Michigan. How did he come to own it? "I got it as a result of something smart I did in 1951," Martin says. "I married the only girl in the world who'd buy me one with her own hard-earned money." He has owned it since 1971, keeps it in the garage, and takes it out half a dozen times a year. The rig was discovered in a junkyard, one of five in various states of disrepair. "I've been able to find any part I need from friends in SPAAMFAA, off of old junker rigs," Martin says.

Post-World War II Rigs

Beginning in 1947, American LaFrance began offering their Series 700 in many variations, all of which seemed to have been successful. Many of these are now out of service and in private ownership. While old enough to be interesting and attractive, they are not so old and precious that they can't be driven for fun (if you can afford the gas). They are out of the expensive realm of the serious collector—and since there are so many of them, likely to remain that way.

Even more affordable and accessible are the Series 900 rigs just coming out of service as reserve apparatus. Zoltan Szucs, an avid collector, was eight years old when he visited the neighborhood firehouse and climbed, under supervision, on a

San Jose's 1931 Mack is remarkably well-preserved: it is still in excellent running condition and wears its original paint and gold leaf. You could fight a fire with it today, although the days when it carried half a dozen firemen, four on the tail board, are long gone. Far left, San Jose's 1931 Mack serves the San Jose Fire Department Muster Team as one of several operational rigs.

1963 American LaFrance triple-combination pumper, a Series 900. He later learned to drive a stick shift on this same rig, became a volunteer, and then a reserve member of the department that used it. When it finally came up for sale after thirty years in service, Szucs bought it and parked it in his driveway. He got it, in perfect condition, for a $2,000 bid. Now he is building a new home for himself and his rig (designed as a firehouse with room for the engine and his collection of antique helmets, nozzles, extinguishers and related material).

The Mack, after over sixty years in front-line service, reserve, and muster team duties is starting to show a little wear. It has developed a hole in its radiator and the oil pan needs replacement, both attended to by firefighters on their off-duty time. Left, Mack has long been popular with the Fire Department of New York, with hundreds of sales over the years. Far left, This 1950-vintage Mack has a handsome, FDNY paint scheme.

149

Mack

Mack is a truck builder whose products included a line of fire apparatus for many years, although it has been recently dropped. There are many Mack rigs in private ownership, most are in the East. Although the marque lacks some of the gaudy glamour of Ahrens-Fox, Mack has a reputation and a following because of Mack's commitment to solid reliability—a prime virtue of this particular breed. According to Don Wisinski, who worked on just about everything during his years as a master mechanic and battalion chief, Macks always arrived with the fewest "bugs" and were the easiest for the firefighters to transition to. When asked if he had a favorite manufacturer, Wisinski replied, "It would have to be Mack, just because of their dependability."

The company was started in Brooklyn, New York, with two Mack brothers (Augustus and Jack) building wagons back in the 1880s. In 1901, Jack got a ride in an automobile—and came up with the idea of combining a gasoline engine with a large, commercial chassis. The idea materialized two years later in the form of a bus. While the company kept making horse-drawn wagons, they also tapped into the rapidly developing automotive market with another bus. By 1904, Mack was building their own engines.

Mack was—and is—a truck builder, first and foremost. Other fire apparatus builders may start with the pump and build the rig around it, but Mack started from the ground and worked up, using a solid truck foundation for the rest of the rig. Their first fire truck came along in 1911, a pump attached to a standard truck chassis.

When the small farming town of Salinas, California, decided to sell off its three 1953 Mack L95s after nearly forty years of service to the city, Cliff Smith knew he had to have one. Below, the L95's engine and trademarked Mack bulldog. Opposite page, left, Mack introduced a whole new line of designs in the 1930s and promoted them heavily in ads like this one from *Fire Engineering*. Mack promoted the quality of their engineering over economy or technological bells and whistles. Opposite page, right, another one of the very popular Macks from the 1950s, this one making an appearance at the annual Fireman's Muster held at Redlands, California.

The Mack company made an important contribution to the gasoline-powered apparatus market in 1915 with the introduction of the famous, and now collectible, Model AC Bulldog truck. About 5,000 were built, many for military service. This vehicle was used for everything and anything

151

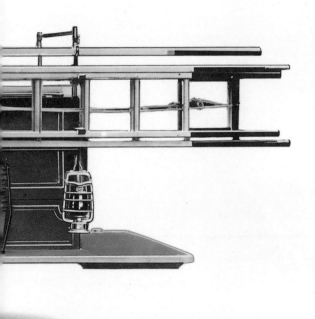

before and during World War I. While most other vehicles of the type at this time were notoriously unreliable, the Mack, with its powerful four-cylinder engine, helped validate the concept of automotive power.

There are not many ACs left. The Fire Museum of Maryland has one, and so does the Hall of Flame. San Jose's muster team has one, too, tucked away in a warehouse, covered with dust. It may have been reliable, but it is an ugly design. Many Macks from the 1920s and 1930s are still around, however, including the APs, the Series 50, 60, and 70 trickled out as pumpers, combinations, aerial ladders, rescue wagons, and all the other permutations. Chicago, among other departments, bought dozens.

These Macks, in and out of the fire service, were such a hit with their owners that they began to get the bulldog nickname around the time of World War I, a tribute to the tenacity of the breed. In 1922, the company adopted the now-famous symbol as their trademark and kept it, and the reputation for reliability, ever since.

A variation on the wartime AC came along in the 1920s. The rig looked nearly identical to the AC except that inside it had a beefy six-cylinder

A 1928 Prospect Master Fire Fighter. Little rigs like this were produced by many small shops in and around Detroit, Michigan, home of Ford, Chevrolet, and the other major manufacturers, each competing to come up with a simple, economical, all-purpose fire engine for small, usually rural jurisdictions. A few rigs like this survive, but not many. *Hall of Flame collection.* Below, the 1927 Pirsch All Power rig from the Hall of Flame collection.

56

The cockpit for the Pirsch aerial was assembled from a wide variety of off-the-shelf parts from many truck builders and suppliers. Below, a small hydraulic pump running off a drive shaft from the differential provides the power to elevate the massive ladder assembly. Far right, Seagrave engines like this 1921 model are now quite rare and valuable.

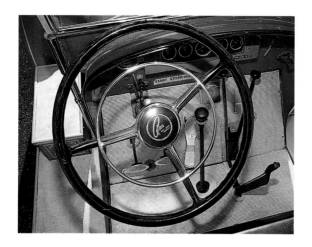

engine. One of these was purchased by the Seattle Fire Department and was delivered in 1927. The rig went from the factory to Berkeley, California, for the installation of its 1,000gpm pump unit, a series-parallel multistage centrifugal made by Byron-Jackson. The trip to Seattle, 713 miles, took three days and twenty-four hours of driving time—an average of 29mph. It was considered a record at the time and a tremendous demonstration of speed and power. According to a glowing newspaper account at the time, "it was the longest and fastest run ever made by a piece of fire apparatus."

When demonstrated for a multitude of fire chiefs observing the engine's acceptance tests, the

1,000gpm pumper actually delivered 1,973gpm at a sustained pressure of 120psi while the six-cylinder powerplant ticked along at 1600rpm. It was an impressive performance, even for the die-hard horse community.

After World War II, Mack sold a popular series of rigs: the 500gpm Type 45 fire engine with full cab, the 750gpm Type 75 with the traditional open cab, the Type 85 with closed cab and 750gpm pump, and finally a 1,000gpm pumper called the Type 95. These sold well and departments like Chicago, long a member of the Mack fan club, bought fleets of the rigs including thirty of the Type 95s in the middle 1950s.

The Series B came along in 1954 with the latest thing in powerplants, a hefty six-banger thermodyne gasoline-fueled engine; a diesel version of the Thermodyne was being implanted after 1959. The Mack B was a successful design and stayed in production for thirteen years—a rather long run for fire apparatus designs.

In 1956, the Mack Trucks firm purchased what was left of the old Ahrens-Fox company. The new owners pried the nameplates off Ahrens-Fox's last design, an excellent and innovative fire engine with a cab-forward configuration. They bolted on new nameplates and introduced the rig as the new, improved Series C from Mack Trucks. This was another hot item with the Mack sales department to the probable distress of the former owners and designers. Perhaps some of the success had to do more with Mack's proven reliability than the excellence of the basic design, but for whatever reason the C sold like proverbial hot cakes.

The Fire Department of New York—the Mecca for fire apparatus sales staff everywhere—liked the C at first sight and bought them by the dozen. The FDNY's first automatic transmissions came surrounded by Mack Series C sheet metal and delighted the chauffeurs (as the drivers are known in New York) who had to negotiate city traffic. When the FDNY contracted with Mack to modernize the whole department, largely as a consequence of these former Ahrens-Fox designs, the former Ahrens-Fox Company owners must have despaired. New York bought 100 Mack C rigs in 1958 and 1959.

Mack's C design looked suspiciously like the American LaFrance equivalent of the time, the

1926 Seagrave Suburbanites are another desirable collectable rig. This one is still owned by the Morgan Hill Fire Department. Right, Seagraves lack the garish glamour of the Stutz and Ahrens-Fox rigs, but Seagraves are often more affordable to buy and operate, so hundreds of them are in private collections all over the United States and Canada.

Series 700, and the two companies sparred for sales. Both were selling in large volume. Mack stopped selling its own line of fire engines in 1990 but still sells chassis to other builders.

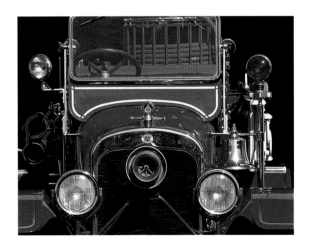

SPAAMFAA members own a *lot* of Macks, from 1920 ACs to pumpers from the 1960s, with rigs from the glorious 1940s predominating.

Pirsch & Sons

Peter Pirsch & Sons is another old apparatus builder. The company began in 1857, building carriages and wagons. This line of business expanded to hook and ladder rigs, and in 1916 into motorized fire apparatus. Initially, Pirsch was just one of the many companies trying to cash in on the big market that had developed to convert the fire service from steam and horse power to automotive. The first rig, a triple combination, was built on a White Motor Company foundation. A 1,000gpm pumper would come along in 1929.

Pirsch is known for a wide variety of models and body styles. The reason for this variety is less

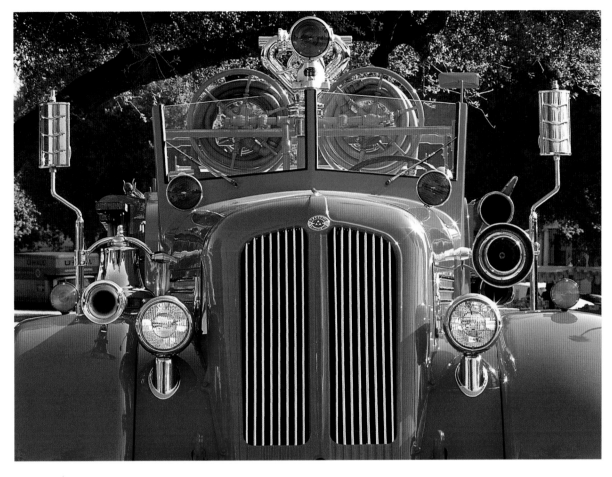

Another Seagrave puts in an appearance at a well-attended muster. Such events are especially popular in the Northeast, where most collectors and rigs are. Musters are great opportunities to view historic rigs of all eras, some of which are in pristine condition and others that don't even run. Musters attract collectors of vehicles, helmets, accessories, plus the folks who enjoy watching the old engines pump. Below, here's what a mid-1920s Seagrave Suburbanite looked like fresh from the factory. Virtually all the builders had teams of specialist painters who did nothing but pin-striping, ornate lettering, and sometimes—as on this rig—scene painting. *Juan Diaz collection*

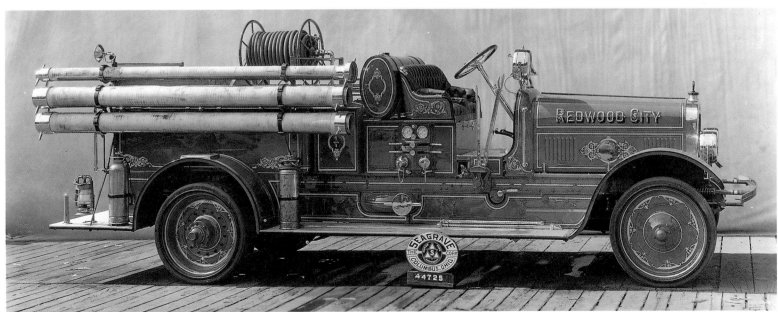

The Seagrave logo on Staunton, Virginia's, 1938 aerial ladder. This vehicle is part of the Hall of Flame collection. Right, fire engines, like people, die in different ways. This is another 1926 Seagrave Suburbanite, rusting and rotting in a field behind a fire station where it once served. Although it is not entirely beyond resurrection, many essential components have been stripped and it is probably beyond salvage.

well known, but is revealed by Dana Martin. Martin may own an American LaFrance but he worked for Pirsch, back in the early 1940s. "It was a very unsophisticated manufacturing operation," Martin says, "and they loved it that way. When I worked there in the summer of 1943 we had one 60-inch sheet metal break—and that was the most expensive piece of equipment we had in the whole shop! Aside from a couple of ladder jigs that were made in the shop, the rest of it was crescent wrenches and hammers. Pirsch would do their utmost to create something out of sheet metal and bar stock and make it look like it came out of tools and dies. There was no machine shop at all; that kind of work was all jobbed out."

Engines and chassis were purchased from other sources, often with a casual attitude about consistency. That meant, according to Martin, that a department could order a Pirsch model and get an engine that might come from any one of four or five manufacturers. Pirsch did not seem to mind, as long as the powerplants had about the same displacement and torque specifications. Any

of these engines might be bolted to the standard (to the extent there was a standard) clutch, transmission, and drive line. "If you didn't want a Waukesha engine and preferred a Hercules," Martin recalls, "Pirsch could build the rig for you that way. He might even slip a Continental in there, if a Continental dealer happened to be on the selection board of the buyer village."

Some observers have commented on Pirsch's innovative addition of an enclosed cab for post-1940 rigs. Those enclosed cabs, Martin explains, were bought from General Motors Company (GMC) in bulk and grafted onto the Pirsch body—or a GMC body, if one was handy and willing. While giving the firefighters some protection from the elements, these cabs were never intended to accommodate people wearing high fire helmets.

Pirsch rigs came with much variety in pump design and capacity. This appears to have been more a result of what was available at the time rather than anything premeditated. As Martin describes, the production floor in 1943, "There

were usually about three times as many rigs in production as they hoped to finish within the foreseeable future and production was entirely dictated by what Railway Express delivered each day A carburetor might show up for one, a set of shackle bolts for another. There weren't nearly enough mechanics to work on all the rigs, so we'd just work on the ones we had parts for." Martin's job was to carry around a can of paint and a brush, following the mechanics and fixing the inevitable chips in the finish that occurred during final assembly.

Despite the trials and tribulations of attempting industrial production during the World War II years, Pirsch actually produced many superb rigs and earned a loyal following. Chicago bought many pumpers and aerials over the years. After the war the Pirsch company designed and built some

superb ladder rigs, and prospered before going under a few years ago.

Pirsch is well-represented in the SPAAMFAA rolls, with dozens of vehicles listed, from a 1924 model A pumper/ladder rig on an Oshkosh chassis to a 1960 Model 41B pumper built on one of Pirsch's own chassis.

Seagrave Company

Although American LaFrance certainly has been the most famous of apparatus builders with the general public, among the fire service, the Seagrave Company has been traditionally a respected and admired innovator and builder of quality rigs.

The company began in 1881, in Detroit, as a manufacturer of ladders for use in local apple orchards. Frederick Seagrave must have made good ones because the local fire departments bought

Among the easiest to buy and most difficult to own rigs are the aerial ladders. They are difficult to drive and worse to park. In fact there's very little you can do with one, unlike a fire engine that can usually accommodate a whole neighborhood's supply of kids for a trip to the ice cream store. That hasn't stopped Leonard Williams, though, who bought this 1941 Seagrave to add to his collection of six other fire engines. Only one fits in his garage at home; the rest are stored in a warehouse.

This page and previous pages, Williams takes the big Seagrave out for a spin with the help of a friend, a qualified tiller operator. Driving the back end of these things is really tricky; when the rig makes a left turn, the tillerman has to know how and when to steer—first to the right and then to the left.

It had a 4-inch straight exhaust pipe—none of this muffler stuff! Well, we had an underpass on West Julian Street, and whenever we'd go through there I'd always gun the engine—the blast of noise in that little tunnel would just about blow the guy riding on the tailboard right off the rig!

—Loren Gray, retired fire engineer

162

many. One of these departments required a small, two-wheeled wagon to transport the ladders to fires and Seagrave built it. There was enough of a demand for the little hand-drawn rig that Seagrave started building many of them. It appeared to be a substantial market so the firm designed and built a nice hook and ladder rig with four wheels.

Spring-Assisted Ladder

The 1890s were a period of rapid urbanization, and in the fire apparatus industry, an era of improved ladders and water towers to cope with the tall buildings that filled up Chicago, New York, and smaller cities around the country. American LaFrance manufactured an 85-foot aerial, but it needed to be erected by hand and took at least four, sometimes six, men to lift it. Seagrave came up with a mechanical assist for elevating ladders and got a patent on it in 1901. The system used coil springs for the first phase and a hand-cranked screw drive to complete the process. While a bit complicated, it was an effective solution to the problem and an immediate success. Other manufacturers soon used similar systems.

Seagrave sold their first motorized rigs in 1907 to the Vancouver, British Columbia, Fire Department—a chemical truck and two hose wagons—all powered by four-cylinder, air-cooled engines.

Ladders continued to be an important part of the Seagrave product line and continued to be an area of innovation for the company. In 1935, the firm introduced the first all-welded alloy-steel ladder. Seagrave came out with the first fully hydraulic elevation system at this time, too. In 1938, Seagrave added safety locks to the ladder sections that prevented inadvertent retraction—a problem that occurred with earlier systems.

Centrifugal Pump

Seagrave's first engines were built about 1900. In 1912, the company developed a completely new form of pump, the centrifugal. Earlier piston and rotary-gear pumps could deliver tremendous quantities of water but had the nasty problem of no convenient shut-off capability. If somebody accidentally shut the water off at the nozzle, the hose would certainly burst. If the pump outlets were shut while a piston or rotary pump was operating, then large metal components bent or broke.

The Seagrave's centrifugal pump prevented these accidents while delivering just as much water pressure and volume.

The centrifugal was an ideal match to the speed of the gasoline engine. It tolerated large quantities of sand and gravel in the water flow and was a simple, relatively maintenance-free design. Centrifugals actually wear in rather than wear out as the sand and gravel in hydrant water polish the inside of these pumps, reducing friction and improving performance. It is still the predominant design today.

Also in 1912, Seagrave offered another important and durable development, the pump pressure regulator. One of the prime functions of an engine is to deliver a steady supply of water at a constant pressure and volume. Prior to the Seagrave regulator, the fire engineer had to manually adjust the pump controls to maintain constant pressure.

Seagrave added another innovation in 1915 with a system of auxiliary engine cooling. Prior to this time, engines tended to overheat while standing by the curb and operating at high rpm. The Seagrave system used hydrant water to help cool the engine, improving the engine's performance and reliability. Seagrave was the first to use hydrant water for indirect engine cooling. American

Buffalo Fire Appliance Corp. sent this fire engine about as far west as any of their rigs, to Ironton, Ohio. Buffalo had a kind of distinctive, sleek style—more car-like than most designs of its day.

LaFrance just diverted some of the hydrant water directly into the radiator, along with lots of the inevitable sand and gravel that comes out of the hydrant. The Seagrave solution was quite sophisticated, a kind of heat exchanger. Ahrens-Fox had a similar system.

V–12 Engine

Until the 1930s, Seagrave, like the competition, relied on big, simple four- and six-cylinder powerplants, and these were generally suitable for the smaller, slower rigs of the early years of automotive-powered apparatus. But by the 1930s, buyers were expecting more speed and agility; Seagrave went to V–12 engines for the additional power needed to propel the heavy fire apparatus quickly through the streets. The Seagrave version was a modified Pierce-Arrow design, a small-block version adapted to the needs of a fire engine's engine. This same basic powerplant was used from 1935 until 1950 with minimal changes.

Styling

Seagrave rigs went through several major style changes between the beginning of the motorized era and about 1950s, from sharp, angular, slab-sided designs, to rigs with a sleeker, more aerodynamic look—and back to the big, sharp, angular look again in the 1970s. The first were huge, boxy contraptions with pumps installed by the Gorham Engineering Company of Oakland, California, in a cooperative marketing effort. Initially, Seagrave built the chassis and shipped them to Oakland for installation and testing of the pump, but the additional shipping costs inhibited this policy, and most of the forty-one Seagrave-Gorham fire engines were assembled in Seagrave's Columbus, Ohio, plant. These fire engines, built between 1912 and 1915, looked like locomotives with rubber-tired wheels.

Then, beginning in 1915, a much more attractive rig hit the streets. It was built around F-4, F-6 and H-6 powerplants and new centrifugal pumps of 350 to 1,250gpm capacity.

Starting about 1921, the Seagrave Company introduced another generation of motorized fire engines and trucks that had a distinctive, stylish quality that hasn't lost its appeal through the decades. These vehicles all incorporate a handsome, rounded cowling—a large aluminum casting—that are beautifully proportioned. The hood and grille are likewise rounded in well-proportioned ways. The whole package gives the impression of a muscular, agile, hard-working truck, ready to go places and do things.

The rigs came in three basic chassis sizes—the smallest, the Suburbanite, was introduced in 1923; typically, a 70hp Continental engine powered it, frequently attached to a 350gpm pump. New York City bought fifty Suburbanites, but without pumps; they were used as hose wagons. Other cities used the Suburbanite extensively.

One of the most durable Seagrave designs was the "Anniversary" series, introduced for the company's seventieth birthday and in production for nearly twenty years. This distinctive design used an engine-forward configuration and has the siren planted dead-center in the cowl, making identification on the fire ground extremely easy.

Seagrave is still in business, cranking out big, boxy fire engines and ladder rigs. It is one of the most popular makes with collectors, with many dozens of vehicles on the SPAAMFAA rolls—from a 1911 Model D chemical hose truck to a 1966 Anniversary series 1,000gpm pumper.

Other Builders

Buffalo Fire Appliance

Buffalo Fire Appliance was once one of the more popular builders, and is still one of the most numerous names on the roster of rigs in private ownership. Despite the popularity and numbers, Buffalo is virtually unknown west of the Mississippi and nearly all of its surviving vehicles are along the Eastern Seaboard.

The company was founded in Buffalo, New York, in 1920 with the intention of getting into the business of helping all those cities convert from horse-drawn steamers to automotive-powered pumpers. Buffalo, like the Howe Fire Apparatus Company and others, relied on truck chassis from other suppliers, to which they added their own components for most of their products. Later, just in time for the Depression, Buffalo introduced their own chassis.

Buffalo rigs have a sleek, 1940s look to them, and a distinctive style. Even though the company failed in 1948, many rigs can be found in the communities of upper New York state.

Four Wheel Drive Auto Company

The Four Wheel Drive Auto Company (FWD) was a builder of specialized, rough-terrain trucks. The firm dabbled a bit in the fire apparatus market before being swallowed up by an industry consolidation. New York bought a few of these FWD rigs in the late 1920s.

FWD unloaded some of the last spring-powered wooden aerial ladders on the FDNY in 1955, a whopping order of over two dozen rigs. By this time, most departments had noticed that wooden ladders burned easily and that the all-metal ones did not. Departments also noticed that hydraulic hoists were far better than the old spring-loaded erection set, but perhaps there was a traditionalist in the FDNY purchasing department.

FWD sold a compact, metal ladder from the Dutch company Geesink, which was a better ladder than the wooden design. During the early 1950s, the new ladder competed with the Maxim Motor Company's German Magirus. Both put the turntable at the back of the vehicle and used more sections than typical American rigs of the time, resulting in a compact little truck with an extended reach.

In 1963, FWD bought the Seagrave Company (and the residue from Seagrave's earlier acquisition of Maxim) outright. Seagrave continued to operate under the old name, however, and FWD stuck to rough-terrain truck manufacture.

FWD is one of the more obscure builders on the SPAAMFAA roster, with only about twenty rigs listed. There's a little bit of everything on the list, though, from a 1926 750gpm pumper to a 1958 airport crash truck.

Howe Fire Apparatus Company

The Howe Fire Apparatus Company is one of those builders whose name has slipped into obscurity from something resembling fame, and that is a shame because this company made a lot of good fire engines. Howe started business in Anderson, Indiana, in 1872, just a year after the great fire that consumed most of Chicago inspired communities to take fire protection a little more seriously.

Buffalo was a regional apparatus builder serving the Northeastern United States from its Buffalo, New York, home. Despite occasional ads in the professional magazines like *Fire Engineering*, Buffalo sold few rigs in other parts of the country. The surviving engines tend to still be in the Northeast, and the trademark is nearly unknown elsewhere. A surprising number of Buffalo rigs are owned by SPAAMFAA members.

The company offered a new, improved, patented piston pumper. In the 1880s and 1890s, the firm also offered hose carts and ladder rigs, in addition to their hand pumpers. As automotive power became a factor in the market, Howe tried selling horse-drawn, gasoline-powered pumps.

One history of the Howe company claims that Howe may have achieved priority with a rig sold to La Rue, Ohio, in January 1906. This engine used a gasoline engine to power a 250gpm pump—but left the hand-operated engine brake arms attached as a backup.

While other builders, particularly the big ones, designed and built their own chassis and sometimes engines, Howe was one of several that used existing commercial chassis and built a fire engine around it. Ford's Model T and Model A were the foundations for hundreds of Howe fire trucks and engines, many still in existence.

The Defender series was offered during the 1930s, a pumper built on a Defiance truck chassis. The line stayed in production for decades; in 1953 the Defenders were redesigned with a bigger chassis, big pumps (750, 1,000, and 1,250gpm capacities) and a Waukesha six-cylinder engine. Another redesign was completed in 1960 when Howe converted the Defender to a cab-forward model that could accommodate five firefighters. Many of these Defenders are still on the street and in collections.

Howe used many brands of commercial truck chassis, including REO, Federal, Diamond T, Nash, Studebaker, Ford, Chevy, and Autocar. The company had a good reputation for innovative design and reliable, efficient construction through the post-World War II years. Howe pioneered pump-and-roll technology for fighting fires from a moving vehicle, along with other refinements that made the business of the engine company more efficient and effective.

In 1961, Howe bought up the Oren Nash Fire Apparatus Company in yet another of the consolidations of the industry—only to be bought up itself fifteen years later in 1976 by what was then the Grumman Emergency Products Corporation. Grumman relocated the company to Roanoke, Virginia, and phased the Howe name out over the next few years. By 1980 this excellent company had become another casualty.

Howe built little rigs on little truck chassis—easy to park, easy to maintain truck chassis—and that's probably why there are plenty still around. Visit a firefighter's muster anywhere in the United States and you'll probably find a cute little REO Speedwagon, Chevrolet, Diamond T, Ford, or White truck foundation embellished with pumps, hose, ladders, and blazing red paint. SPAAMFAA lists several dozen, from a 1914 rig that was originally horse-drawn, to a 1960 pumper built on a Dodge frame and powerplant.

Maxim Motor Company

Carlton Maxim founded the Maxim Motor Company in 1888 in Middleboro, Massachusetts. The company began building fire apparatus in 1914. Like the many other vehicle builders of the time, Maxim got into the new market with an only slightly modified version of their standard truck—a chemical hose wagon. A bona fide pumper hit the street in 1915 and improved versions followed during the 1920s.

Among this company's few claims to fame was the introduction of the Magirus, the German compact aerial ladder. These aerials were just the ticket for the confined, twisting European cities and a few were sold in the United States. It was an idea a bit ahead of its time, though. Even so, the Mack Company bought Maxim ladders for their own rigs, a testament to the quality of their design.

Maxim lasted quite awhile before getting sucked into the Seagrave Company in 1956. For the next seven years, production of Maxim apparatus continued independently, but when Seagrave itself was absorbed by FWD in 1963, the Maxim name disappeared.

W. S. Nott Company

The W. S. Nott Company was a popular manufacturer of steam fire engines during the late nineteenth century and was once one of the leading names in the industry. The company converted its apparatus to gasoline power in 1912 and was soon selling big, handsome pumpers with an interesting worm-gear drive train.

The company was quite successful in its marketing efforts during this early phase, selling triple-combination engines to the FDNY in 1913 and an innovative hose-wagon/pumper in 1915. The hose-wagon/pumper was an important development because, previously, a response required two or even more units to be dispatched, including a pumper, hose wagon, and, especially during the horse-drawn steamer days, a fuel tender. An engine

company often had twelve full-time members, with two officers and ten firemen; the new combinations like the one from Nott (and from all the other builders about the same time) offered a radical change in the business of urban fire companies, with far smaller staffs and much lower costs.

That Dam Fire Engine Company's 1931 REO Speed Wagon is a good example of a successful restoration effort. The little engine was just another pile of rusting junk when discovered and now its a prize- winning entry in Western firemen's musters.

Stutz didn't make many fire engines but the few that have survived are quite collectable. This attractive radiator cap is attached to a 1924 K2 engine (the company's smallest) that once served the town of Havre de Grace, Maryland, until 1935 and Farmland, Indiana, until the 1950s.

There are six Nott rigs on the 1992 SPAAM-FAA roster, spanning only 1924 to 1931. They are all little pumpers, the end of the line for one of the most famous and respected builders of steam pumpers from the previous century.

Sanford Fire Apparatus Company

Another truck builder with an eye on the fire apparatus market was the Sanford Company. The firm was started in 1908 and built trucks until 1925 when it produced its first home-brewed fire engine. Sanford used its own chassis as well as those of other manufacturers, depending on the specifications of the buyer. Their own 500 series sold well, usually with a Waterous pump installed. But they also converted Ford Model Ts and other commercially available chassis during the pre-World War II years, despite a rapidly declining market.

Sanford's major products during this period were the 528 pumper, a 500gpm fire engine, and the Cub, a cute little critter intended to compete with Seagrave's popular Suburbanite. The Cub was designed for smaller communities that were only just now converting from hand or steam pumpers—or nothing at all—to modern firefighting equipment. This was a considerable market between the wars, usually overlooked because of the glamour of the big city departments with their gaudy Ahrens-Fox or American LaFrance rigs.

Stutz Fire Engine Company

The Stutz Fire Engine Company may not have built many fire engines, but the ones they manufactured are among the best survivors of all antique apparatus. That is probably because of the Stutz name, associated as it always has been with high-performance cars such as the Stutz Bearcat. Dr. Peter Malloy, director of the Hall of Flame, estimates that as many as half of all Stutz rigs ever built are still around, primarily because of the name factor.

The company was founded by Harry Stutz to build motorized vehicles of all types. While the Stutz Bearcat was certainly the most famous, the company started building fire apparatus in 1919 with the delivery of a triple-combination pumper to its hometown of Indianapolis, Indiana. Like other builders of the time, sales tended to be regional; Indianapolis was a good Stutz customer and bought almost three dozen rigs when it modernized in the early 1920s. The company offered a full line of engines and trucks during the 1920s, from little 350gpm units to the big 1,200gpm pumpers. The 350 K model (nicknamed the Baby Stutz) was yet another competitor for the small-town market, with the Seagrave Suburbanite and the Sanford Cub.

Stutz had a reputation for doing everything fast—including going out of business. After a prosperous few years, the company went under in 1928. There was an attempt to resuscitate the carcass in a venture called the New Stutz Fire Engine Company that lasted until 1940. This firm was responsible for the first diesel-powered fire engine in the United States, a 1,000gpm pumper ordered by Columbus, Ohio, in 1939.

Stutz rigs are among the most popular in the collector market, perhaps partly because of the association with the Bearcat. There are about two dozen on the SPAAMFAA roster, nearly all pumpers and dated from 1920 through 1935.

Ward LaFrance

There ought to be a law against having two big companies in the same industry with similar names, but there isn't and that's why there are two fire engine builders with LaFrance in their name. Even worse, both were located in the same small city of Elmira, New York. Although the families were once related, the companies they spawned were not. Ward LaFrance was another truck builder that decided to cash in on the lucrative fire-engine market back in the Depression years of the 1930s with custom-built pumpers. In a time when hardly anything was selling, Ward LaFrance discovered a rewarding profit center for their product.

Ward LaFrance tended toward bigger rigs in later life, all based on their successful commercial truck line. These vehicles all have a distinctive grill pattern and a strong, sturdy form that reeks of power and reliability. After World War II, Ward LaFrance prospered along with most of the rest of the market; the Fire Department of New York bought twenty 750gpm engines in 1946. The company sold many more to departments around the country.

To Ward LaFrance goes the extremely dubious honor of introducing lime-yellow paint for fire engines, on the allegation that safety was enhanced through the improved night visibility of this color. Now, it is true that the color is more visible than red—especially at night—but as far as improving safety there is still considerable doubt. Some firefighters are adamantly opposed to the color, partially on the grounds that if someone can't notice the flashing light bar on top of the rig, air horn blasts, siren, and strobe light flashes, then he or she is too drunk to be assisted by a paint scheme. There has been another study since the first one which states that red with a bold white stripe is even more visible than "slime-green" at night.

While some firefighters believe that a fire engine has to be red, others are much more tolerant. Capt. Dennis Madigan says, "As long as it makes my firefighters even a little bit safer out there, then it's alright with me."

Muster Teams

In the middle of the nineteenth century, firefighters' musters were popular social events for fire company members, who competed against each other without the bloodshed common to the rivalry at working fires. Musters provided an opportunity to test engines and firefighters against each other and attracted large crowds of onlookers.

After many years of neglect, firefighters' musters were resurrected in the 1970s among people in the fire service and people involved in collecting apparatus. The muster teams and their competitions are especially popular in the northeastern states where volunteer departments of small cities and towns located in close proximity generate a bit more rivalry than in the West. The groups that participate in musters are dedicated to preserving and restoring apparatus and using them in competitions.

Some of the teams are limited to members of particular fire departments while others are open to anybody. The ones in the West seem to emphasize the car show approach, with static displays of glittering rigs. In the East, many involve competition in events like the ladder climb and bucket brigade. There are teams in most states and all operate independently.

Although there is much variation from one state association to another, typically there are several musters each year with competition in several basic categories: motorized equipment, hand pumpers, bucket brigade, and hose cart. At the conclusion of these events, there will not be a dry thigh in the house—everybody gets soaked, spectators included. The Eastern version of these proceedings are sometimes called pump-ins and are much wetter than those in the arid regions of the country.

For up-to-date muster information and local contacts, look in the annual listing *The Visiting Fireman.*

Some firefighters are adamantly opposed to the [lime-yellow] color, partially on the grounds that if someone can't notice the flashing light bar on top of the rig, air horn blasts, siren, and strobe light flashes, then he or she is too drunk to be assisted by a paint scheme.

Chapter 6

Modern Apparatus

As you have read, firefighting and the tools to accomplish it have changed tremendously since the end of World War II. One of the most important changes, and least noticed, is that there are fewer fires than there used to be. Education, smoke detectors, better construction materials and methods, and sprinklers in public buildings have contributed to the trend away from major fires in urban areas.

But while the number of fires has diminished, the type and danger from fire emergencies has risen. In the past when firefighters responded to a structure fire, all they had to worry about was wood smoke and flames. With modern buildings full of plastic furniture, appliances, carpets, and large volumes of synthetic materials, fires tend to produce a cloud of toxic chemicals rather than a column of simple smoke. The danger from fire has been largely replaced by the danger from chemical gasses. The result is a different kind of fire engine and firefighter.

Today's fire departments use their rigs for a much wider diversity of emergencies than departments of the past. When the rigs roll out of the station today, it is much more likely to be to a car wreck or a heart attack than a house fire. As a result, modern fire engines and trucks have been adapted to this revised mission with additional equipment. The contemporary fire engine may have about the same pump capacity as the steamer of a hundred years ago, but it can do a hundred things that the steamer and its whole company could not attempt.

Engines

Contemporary fire engines are typically triple combinations, complete with about 500 gallons of water, a 1,250gpm pump, and an assortment of hose for all occasions. In the rig's commodious compartments are light water concentrate (for foaming down fuel spills), forcible-entry tools, and more medical support equipment than was once found on a rescue rig. There is also an assortment of nozzles, a high-capacity deck gun, or a monitor (similar to a deck gun but removable from the engine). Almost every engine carries one or two ground ladders that can get firefighters up on the roof of a single-story structure in a pinch.

Today's engines are about twice as big and heavy as their predecessors. Much of that bulk and mass goes into better protection for the crew. The engines and transmissions in today's fire apparatus are as superbly engineered as they were in the past. A normal diesel powerplant is expected to last through twenty years of service (the worst kind of service, from dead cold to full speed, and back to off again, being the normal routine) without ever having the head removed.

Aboard the engine are spare air bottles, back boards, nozzles, a resuscitator, automated defibrillator to jump-start heart failures, plus the same old 1-inch hand line that has been aboard fire engines since the invention of the chemical wagon.

Trucks

A contemporary fire truck will often have the same gear that the engine company carries, but in

Airport crash rigs are among the most expensive of fire apparatus, with huge water and foam tanks, engines, and additional specialized equipment. This rig belongs to the US Air Force.

Van Pelt delivered many rigs in many configurations to departments around the nation. This one features plenty of room for gear not found on earlier fire engines.

a different mix. Sometimes, the only way to tell the difference is to read what the department has decided to letter on the rig—if it says "truck," then they are using the vehicle for "truckie" work, and a truck it is. You can ignore the pump, hose, and deck gun.

Rescue

Specialized rescue vehicles have long been a fixture of fire departments, but you have to be careful about what you call these rigs. In some parts of the country, they are squads while in others (the next township, for instance) the same rig is

the rescue unit. Regardless, these trucks typically carry all the heavy rescue gear that can extract you from the inside of a twisted ball of steel that once was your car. That includes the Hurst (and similar) "jaws of life" and the associated cutter that can pop open a mangled door, cut a thick steel window post, or slice the roof off a truck or train car. Also aboard are air jacks that can lift just about anything, plus a vast assortment of wooden blocks and wedges to keep it lifted.

HAZMAT

An amazing assortment of materials that burn, explode, radiate, fume, and corrode travel the roads of every city. Inevitably, they bump into things and leak. When that happens, it is time to call in the specialists, the hazardous materials (HAZMAT) team and their big rig. Aboard are suits for all occasions—acid spills, toxic fumes, PCBs, infectious wastes—each with its own hazards and requirements. The HAZMAT rig usually has a computer, reference library, wind speed and direction instruments, and decontamination equipment—but no hose, tank, not even a single ladder.

Aerials

Aerial ladders haven't grown much in the last fifty years and are still about 100-feet long. Some are reverting back to wooden construction, partially because wood is an excellent insulator, partially because aluminum, steel, and fiberglass all have problems that make them less than ideal for fire service. The original argument against wood was that it was difficult to maintain, but that has changed with new impregnation processes that provide lifetime protection. And aluminum, for all its virtues, is subject to metal fatigue cracks that can lead to failures. Fiberglass is a great insulator but is easily damaged by sunlight and exposure to heat.

And, while the old tillered aerials were once thought to be involved in more traffic accidents than straight ladder rigs, there is a renewal of the design. The tillered rigs will go around corners no conventional aerial can attempt.

Airport Crash Trucks

The biggest, most expensive rigs in any department today are likely to be the airport crash trucks. These monsters confront aircraft fires, one

Rural departments often opt for less elaborate and less expensive commercial truck chassis adapted to fire service use. This one normally serves the small Oregon community of Lebanon with its big water tank and front-mounted pump.

173

A 1974 American LaFrance 100-foot aerial with ladder pipe in action at a five-alarm call in Cleveland, Ohio, on June 11, 1979. The outriggers allow the ladder to be fully extended. *John McCown*

of the most challenging missions in firefighting. An airplane fire is a structural fire, but instead of a house or office building, perhaps 200 people are stuffed into a sheet metal structure along with several thousand gallons of kerosene.

A crash truck is essentially an overinflated fire engine, with many extras. One of the most popular is the Oshkosh T-3000, found at airports nationwide. It is 36-feet long, 12-feet high, and

costs a third of a million dollars. It has six-wheel drive, a 540hp engine, and a top speed of 65mph—but with 3,000 gallons of water aboard, it starts and stops on something a lot larger than a dime. It can fight a fire without anybody getting out of the rig through the use of remote controlled nozzles for fog and foam. Once the fuel fire around the aircraft is knocked down, the rig can drive right up to the fuselage to conduct an interior attack on the fire inside, if any.

A special appliance called a Snozzle is on the end of a boom; the firefighters use it to punch through the side of the aircraft and then spray a fine fog of water on the inside, cooling and quenching any flames. Halon is also available on the T-3000, a 500-pound tank with a hose reel and special nozzles. The rig also carries special drills, saws, and rescue tools like the Hurst spreader and cutter, plus air jacks and perhaps the Ajax air chisel.

A 1980 American LaFrance 85-foot snorkel shows its versatility at a 1981 Shaker Springs, Ohio, theater fire. *John McCown.* Below, engines have always been designed to meet the kind of conditions they're expected to face. This classic brush-fire rig is doing what it was born to do.

Cupertino, California's, aerial ladder comes from the Kovach company, a 100-foot aerial ladder with basket from LCI. Far right, this 85-foot snorkel offers many advantages over earlier aerial ladders but it requires a very well trained and talented operator to use. It unfolds in one way only, with the "elbow" coming up first, the basket afterwards. Lower right, the ten firefighters who crew this rig owned by Hicksville, New York's, Citizens Engine Company No. 3 are the latest in a long line of well-equipped volunteers. In fact, volunteer companies often have much more costly and capable engines than their full-time contemporaries, as this rig testifies. For about a third of a million dollars you get tilt-cab convenience, air conditioning, stainless steel construction, 445hp engine, and automatic transmission. Pump capacity is an amazing 1,750gpm and there's 500 gallons of water aboard. Besides the usual discharge outlets, this engine also has one mounted on the front bumper. Hicksville's volunteers maintain an old fire service tradition with this rig by installing their old bell on their best engine. *Sutphen Corporation*

With modern buildings full of plastic furniture, appliances, carpets, and large volumes of synthetic materials, fires tend to produce a cloud of toxic chemicals rather than a column of simple smoke. The danger from fire has been largely replaced by the danger from chemical gasses. The result is a different kind of fire engine and firefighter.

When brush and forest fires get going during the summer fire season across the western United States the "mutual aid" call goes out and rigs assemble from many miles around. Left, a pair of Cleveland's 1979 American LaFrance rigs on scene at a very smoky four alarm fire. *John McCown.* Far left, a fleet of huge water tenders and brush-fire engines assemble at the edge of a grass fire.

Hamburg, New York's Sutphen 100-foot aerial is powered by a 450hp Detroit Diesel V–8, has a 1,500gpm pump, and cost about $500,000. *Sutphen Corporation.* Left, San Jose's Engine 8, a 1989 Pierce. Far left, Campbell's Engine 1 receives a good washing after a fire.

Campbell's Engine 2 in handsome white livery. The rig is a 1991 Kovach triple combination pumper, propelled by a Detroit diesel engine and Allison transmission combination. The 1,500gpm pump comes from the Hale company. Lower right, Campbell, California's, Station 1 A-shift responds to a five-square-mile jurisdiction with a rescue rig and an engine. Far right, firefighters respond far more frequently to calls like this one, a car roll-over, than to actual fires. In response, the rigs now include space for large quantities of emergency medical equipment and extrication tools like the "jaws of life."

U pper Arlington's rescue rig includes many of the functions of an advanced fire engine with all the tools and accessories for rescue calls, including stokes litters, air jacks, saws, cutters, and the "jaws of life." There's a generator and extensive paramedic supplies. A 12,000-pound winch can be used for vehicle recoveries; the pump is rated at 1,500gpm and 500 gallons of water are carried. There's a "deluge gun" on the top deck—and air conditioned comfort for the crew after a hard day on the line. Lower right, this is Sutphen's technology demonstrator, a 105-foot aerial on a special "fifth wheel" tractor-trailer. The ladder stows into an aluminum housing below the tiller cab at the back of the rig. Such tillered rigs were once considered obsolete but are being sold by several builders to large metropolitan departments including Baltimore, Columbus, and New York. *Sutphen Corporation.* Far right, Rescue 1's crew mounts up for another run; they average about ten a day—car wrecks, coronaries, with an occasional shooting or stabbing for variety. Rescue 1 is a product of Southern Ambulance Company, based on a chassis from International. The body is a low-profile type, with room for several victims and paramedics; included are advanced treatment facilities and communications systems.

Chapter 7

Overhaul and Take-Up

Like the automotive industry, the builders of fire apparatus have had a stormy history. Most of the major manufacturers of fire engines have gone out of business, been absorbed into other companies, or have been displaced within the market by newer, more innovative builders.

American LaFrance is still in business, but after a tiff with its unions in the 1980s, the company moved from its ancestral home in Elmira, New York, to Bluefield, Virginia. The professional fire service is almost 100 percent union and this development may have something to do with the sudden drop in orders for the company that has always been the foundation of American fire-apparatus design and construction.

After a few years of trading on the name and its reputation, and floundering around with a variety of non-fire-service markets, American LaFrance has restructured and has refocused its efforts on building fire engines, trucks, and associated rigs. The Virginia factory discarded the old tooling and now uses computer- and laser-aided fabrication technologies. Stainless steel is now the material of choice, and the company cranks out about sixty rigs a year.

A new company called Emergency One was formed in the middle 1970s and has rapidly taken over the market for fire apparatus in the United States. E-One (as it is universally known) is now the biggest builder of trucks, engines, and emergency vehicles in the country. E-One took over the market by emphasizing modular construction that

makes for much more efficient, rapid manufacture than previous fabrication techniques. The company has also pioneered the use of aluminum in chassis construction, which is a good thing because the trend for all manufacturers has been to rigs that are *big*, with many lockers for accessory equipment; lighter weight and corrosion resistance become important factors as the size of the vehicle increases.

Ahrens-Fox, Pirsch, and Pierce-Arrow have disappeared and others, including Mack, have gone back to building plain commercial trucks instead of catering to the specialized, custom-fire-apparatus market.

But if the old reliables have faded away, a new crop of innovators has arrived to take over. E-One is only the leader of a whole pack of new, innovative, imaginative companies that have taken the business over from the old guard. Other leading companies are Grumman, FMC, Oshkosh, Stuphen, Walter, Van Pelt, and the Canadian firm of Thibault.

The modern rigs are huge, impressive, superbly capable on the fire ground and at emergency calls, but somehow lacking the sex appeal of the smaller engines and trucks of an earlier era. Even when not painted that evil lime-yellow color, these rigs just do not seem to have the theatrical qualities of the old Mack, Ahrens-Fox, American LaFrance, or Seagrave engines. Someday even the modern rigs will be taken out of service, auctioned off, or stashed in some municipal warehouse, but

They don't make them like this anymore. These lovely red and gold wheels carried an 1880s American steamer to countless fires. Repairing wagon-style wheels is something of a lost art, so the wheels are often the most challenging part of any steamer retoration.

The best part about driving my old rig is that it's an open cab; hitting the foot-pedal switch for the siren when you're going to a fire, listening to it wind up. You have to drive with two hands and two ears, listening to the radio, the captain's instructions, working through the gears, double-clutching every one, feeling that rig come alive in your hands.

—Zoltan Szucs, fire-apparatus collector

not many of them are likely to be adopted by adoring fire buffs of the future because the rigs of today scarcely fit in the firehouse, much less in a standard garage.

Although it is now past, there was a golden age in the design and construction of fire engines and it lasted from about 1915 to about 1955. It was an age of a more elegant design for a less complicated mission, for rigs that were designed to put out fires instead of cope with a broad spectrum of social ills and disasters. So—were those the good old days?

The old steamer may have been a mechanical marvel, with pump capacity and response times as good as those of today's engines, but it was just an engine and needed the services of several other fire companies and about fifty firefighters to do what a much smaller volunteer or professional company today does with five or ten people and much more adaptable, capable equipment.

Collector Zoltan Szucs appreciates the quality of the equipment from the past. The following paragraphs summarize his feelings.

"The old rigs, from the twenties on up to the sixties, are still around for a reason. They were built by people who'd apprenticed as machinists, who learned to craft pumps from natural materials like steel and bronze, that can work as well today as when they were new. The pump on my rig was the result of 150 years of American LaFrance experience and tradition; it was designed and built to last, and it still drafts and pumps water as well as it did thirty years ago when it came off the assembly line. Today's pumps aren't always like that; there are alloys and neoprene and Teflon that can fail, as one of our new engines did at a major fire when it had to work for a long time and it overheated.

"Many of the old traditions are coming back—tiller aerial ladder trucks, for instance, like the old ones, are being bought again because they turn out to be more maneuverable. Wooden ladders are coming back because aluminum conducts electricity and wood doesn't. And leather helmets, too!

"The older rigs didn't have power steering or any of the high-tech stuff that lets you hop into a modern rig and in a moment you're out the door. The old rigs take some thinking and a little time— it's like getting a dinosaur to come alive—and sometimes I find myself talking to it, praising or cursing it. But once you get it started, it's so dependable!

"The best part about driving my old rig is that it's an open cab; hitting the foot-pedal switch for the siren when you're going to a fire, listening to it wind up. You have to drive with two hands and two ears, listening to the radio, the captain's instructions, working through the gears, double-clutching every one, feeling that rig come alive in your hands. It takes strength, timing, and judgment—not like the new, modern rigs that you can turn with a finger. There's a thrill to feeling the wind in your face, especially when I hit the sirens, turn on all those beautiful lights. You feel it vibrate and shake, all that raw horsepower—and the biggest thrill of all is being able to downshift without hitting a gear. I've been in the fire service seventeen years, and its still exciting. It brings a smile to people's faces when they see an old fire truck. It's not that way with the new ones. The quality of the paint and steel just doesn't match the rigs of the sixties and before.

"The new rigs perform well in many ways, but a little computer chip or solenoid can quit on you and take a rig out of service. A department near here bought one of the best, most elaborate, state-of-the-art aerial ladder rigs with fully automatic hydraulic outriggers and incredible safety devices—but one of the electronic solenoids failed and the rig couldn't retract the stabilizers and was stuck in place. Now, in the 1950s, you just had mechanical, hand-operated screw outriggers and you never have a problem with them.

"The new fire trucks are like jets—you can go screaming down the street in them, but if one little plastic thing goes wrong they're dead. The older rigs, like the steamers and the earlier automotive rigs, might have taken longer to get to the fire, been harder to steer and harder to operate, but they could pump for hours and hours, as long as the coal and the water were there. And many are still pumping today!"

Resources

Recommended Reading

Books

There are many books about fire apparatus and firefighting, and there are bookstores that specialize in nothing but this subject. It is an area that has inspired people to invest much effort in scholarship and the expense of self-publishing.

For Ahrens-Fox fanatics, Ed Hass' two books, *The Dean of Steam Fire Engine Builders* (about Chris Ahrens) and *The Rolls Royce of Fire Engines* (about the Ahrens-Fox piston pumpers), both published by Kes-Print, Incorporated, 1986, include virtually all the detailed history anyone could ever want.

Seagrave fan club members will appreciate Matthew Lee's *Seagrave—A Pictorial History of Seagrave Fire Apparatus*. This book includes many of the superb large-format black-and-white delivery photographs that builders once made to document every rig. Although the text is a bit sparse (it is a pictorial, after all), there is still enough to accurately trace the company's evolution. For books about firefighting, nothing compares to George Hall's two incredible volumes, *FDNY—New York's Bravest!* (with Thomas K. Wanstall) and *Working Fire—The San Francisco Fire Department* (with John Burks), both from Squarebooks. These books provide an exciting insight into life in the fire companies on the West and East coasts, the result of many hours on the fire ground and excellent scholarship. The pictures, text, and design are superb. Both books are currently out of print but sometimes available from fire service bookstores.

Walter McCall's *American Fire Engines Since 1900* is a good pictorial survey by a man who knows his subject. Again, this is not something you will find in your local bookstore, but the fire-service bookstores may have it.

Periodicals

Several magazines cater to the fire service, the most popular of which are *Firehouse, Fire Engineering,* and *Firefighter's News*. All contain much information about professional and volunteer department issues, trade ads, and usually many shots of good burners. Although primarily intended for volunteers and professionals in the fire service, these magazines are entertaining and informative. The classified ads are a great place to shop for books, obscure fire paraphernalia, and that fire engine to park in the garage.

The Visiting Fireman

Visiting Fireman is an annual listing of just about anybody interested in the fire service. For $11 a year, you can find out about musters, collecting, model building, photography, firehouse restaurant reviews, even the frequencies for many of the fire departments in North America and Europe—the definitive reference for fire buffs. To receive this listing, send a check to The Visiting Fireman, 1024 Elizabeth Street, Naperville, IL 60549.

Museums

Dozens of large and small museums around the country (and in Canada, too) attract fire buffs. These museums are sometimes tucked away in odd, difficult to find places, but are usually worth the effort. A complete list is found in the back of *The Visiting Fireman*. (See Recommended Reading above.)

An alternative to the formal museum is often found in your local fire department, back in the corner of a firehouse (or maybe rusting away out back), where departments keep old and out of service apparatus. Drop by your neighborhood station and ask; even if the firefighters on duty don't have anything old to offer, they will often provide a tour of the rigs on the station floor.

Following is a list of the largest and most accessible museums. A call to your local fire department will probably reveal several other collections in museums nearby that might be even better.

County of Los Angeles Fire Department
  Museum
1320 N. Eastern Avenue
Los Angeles, CA
(805) 267-2411

Firefighter's Museum of Nova Scotia

The Firehouse Museum
1572 Columbia Street
San Diego, CA
(619)232-FIRE

The Hall of Flame
6101 East Van Buren
Phoenix, AZ
 (602) ASK-FIRE

The largest collection of fire apparatus in the world in a museum dedicated strictly to the preservation of fire engines and associated equipment.

New York City Fire Department Museum
278 Spring Street
New York, NY 10013
(212)691-1303

San Francisco Fire Department Museum
260 Golden Gate Avenue
San Francisco, CA
(415) 861-8000
Operated by the St. Francis Hook & Ladder Society, this excellent museum has a wide range of materials beautifully displayed. Open only Thursday through Sunday, 1pm to 4pm.

Index

Ahrens, Chris, 131
Ahrens Manufacturing Co., 138
Ahrens-Fox Fire Engine Co., 131-137
Ahrens-Fox fire apparatus, 48, 69, 70, 82, 86, 87, 92, 96, 98, 99, 104, 105, 112, 129, 131-136
Ahrens-Fox Fire Buffs' Association, 129, 132
airport crash rigs, 171, 173
American Fire Engine Co., 51, 69, 138
American LaFrance Co., 137-149
American LaFrance fire apparatus, 22, 24, 29, 51, 60-67, 69, 70, 75, 78, 79, 81, 83, 85, 88, 89, 92, 93, 99, 100, 103-105, 107, 109, 111, 112, 114, 116-118, 120, 122, 127, 129, 131, 132, 137-139, 141, 142, 144, 145, 147, 149, 158, 174, 175, 179
American fire apparatus, 55, 56, 107, 187
ATO Corp., 122
Bechthold, Ken, 52
Berardo, Ralph, 144
booster tank, 95
Buffalo Fire Appliance, 82, 132, 165
Button, 138
Charco, Captain John, 23
chemical weapons, 95
Chevrolet fire apparatus, 85, 93
Chicago Fire Dept., 69, 112, 115, 117
Christie's Front Drive Auto Co., 79
Clapp & Jones fire apparatus, 52, 53, 58, 64, 138
Clet, Battalion Chief Jeff, 16, 21

Crown apparatus, 122
Diamond T fire apparatus, 109
Diaz, Firefighter Juan, 24, 25
Downers Grove Fire Dept., 94
Elmira Union Iron Works, 137
Emergency One (E-One), 122, 187
FMC, 187
Fire Dept. of New York, 31, 65, 70, 95, 116, 149
fire hydrants, 62
Five Wounds Church, 13, 35
Four Wheel Drive Auto Co., 116, 122, 165
Ford, 106, 114
Fox, Charles H., 69, 131
Franklin Fire Society, The, 42
General Motor Co. (GMC), 158
Go-Tract, 122
Gray, Firefighter Loren, 143, 162
Grumman, 187
Hale, Don, 83, 88
Hall of Flame, 43, 88, 130
Hass, Ed, 92, 129, 132, 135, 136
HAZMAT, 173
hook and ladder, 47
hose, 49
I.E.S. Hall, 13
Howe Fire Apparatus Co., 104, 132, 165, 167
Hunneman hand pumper, 45
King, Captain Bob, 23
King, Captain Paul, 16, 17, 18, 21
Knox-Martin fire apparatus, 22, 70-73, 75
Kovach Co., 176, 182
Krapp, George, 131
ladder, 83
    aerial, 95, 109, 173

spring-assisted, 163
Ladder Towers, Inc., 122
LaFrance Fire Engine Co., 138
LaFrance Manufacturing Co., 137
Los Angeles Fire Dept., 112, 140
Mack fire apparatus, 11, 27, 95, 103, 108, 112, 114, 131, 147, 149, 151, 155
Mack Trucks, Inc., 150-156
Madigan, Captain Dennis, 169
Madigan, Pat, 57, 63, 71, 134, 145
Magallon, Firefighter Tony, 16
Malloy, Dr. Peter, 43, 51, 52
Martin, Dana, 147
Maxim Motor Co., 132, 166
Merryweather Co.
Morgan Hill Fire Dept., 9, 97
New Stutz Fire Engine Co., 169
Newsham pumper, 43
"Old 41," 22, 42
Oshkosh, 187
Packard, 82
Pierce-Arrow, 14, 82, 122
Pirsch & Sons, 104, 105, 129, 132, 154, 156
Prospect, 153
pump, 96, 97
    centrifugal, 163
REO Speedwagon, 104, 167
Rumsey hand-drawn pumper, 46
San Francisco Fire Dept., 120
San Jose Fire Dept., 11, 22, 61, 63, 70, 78, 99, 107, 129, 139, 143, 147
Sanford Fire Apparatus Co., 132, 168
Seagrave Co., 159-165
Seagrave fire apparatus, 69, 70, 78,

83, 85, 95, 97, 104, 105, 107, 122, 132, 156, 157, 158, 159, 162, 163, 164, 188
Seattle Fire Dept., 154
self-contained breathing apparatus (SCBA), 16, 34, 35, 118
Shaw, Firefighter Mike, 25
Silsby, 50, 54, 138
snorkel, 115, 116, 117
Society for the Preservation & Appreciation of Antique Motor Fire-Apparatus in America (SPAAM-FAA), 130, 132, 133, 134, 135, 144, 147, 168, 169
Southern Ambulance Co., 184
Souza, Fire Engineer Al, 16, 17
Spangler Dual, 112
Stutz Fire Engine Co., 82, 132, 168
Sullivan, Tim, 63, 65
Szucs, Zoltan, 48, 118, 188
Thibault, 187
Union Fire Company, 50
Van Pelt, 122, 172, 187
Volunteer companies, 41, 42, 46, 48, 50
Walter, 187
Ward, Bob, 134, 135
Ward LaFrance, 169
water tower, 83, 100
Waterous Pump Company, The, 68, 69, 83
Webb Motor Fire Apparatus Co., 69
White Motor Co., 82
Williams, Leonard, 159, 162
Wisinski, Don, 150
Zobrosky, Firefighter Gary, 14, 15, 16